Simple Solutions

Ergonomics for Construction Workers

James T. Albers
NIOSH Division of Applied Research and Technology

Cheryl F. Estill
NIOSH Division of Surveillance, Hazard Evaluations, and Field Studies

U.S. DEPARTMENT OF HEALTH AND HUMAN SERVICES
Public Health Service
Centers for Disease Control and Prevention
National Institute for Occupational Safety and Health
2007

Disclaimers and Ordering Information

This document is in the public domain and may be freely copied or reprinted.

Disclaimers

Mention of any company or product does not constitute endorsement by the National Institute for Occupational Safety and Health (NIOSH). In addition, citations to Web sites external to NIOSH do not constitute NIOSH endorsement of the sponsoring organizations or their programs or products. Furthermore, NIOSH is not responsible for the content of these Web sites.

The views expressed by non-NIOSH authors in these proceedings are not necessarily those of NIOSH.

Ordering Information

To receive documents or other information about occupational safety and health topics, contact NIOSH at:

NIOSH—Publications Dissemination
4676 Columbia Parkway
Cincinnati, OH 45226-1998

Phone: (800) CDC-INFO (232-4636)
TTY: (888) 232-6348
E-mail: cdcinfo@cdc.gov
Website: *www.cdc.gov/niosh*

For a monthly update on news at NIOSH, subscribe to NIOSH eNews by visiting *www.cdc.gov/niosh/eNews*.

NIOSH is a federal government research agency that works to identify the causes of work-related diseases and injuries, evaluate the hazards of new technologies and work practices, and create ways to control these hazards so that workers are protected.

DHHS (NIOSH) Publication No. 2007–122, August 2007.

Acknowledgments

Writing and Research

James T. Albers, NIOSH Division of Applied Research and Technology
Cheryl F. Estill, NIOSH Division of Surveillance, Hazard Evaluations, and Field Studies

Editing and Design

Eugene Darling, Labor Occupational Health Program (LOHP), University of California, Berkeley
Kate Oliver, LOHP
Laura Stock, LOHP
Anne Votaw, NIOSH

Illustrations

Mary Ann Zapalac

Photo Credits

All photos: NIOSH, except: p.23 (bottom) Jennifer Hess; p.27 (both photos) Earl Dotter; p.29 (bottom) Racatac Industries Inc.; p.31 (both photos) Non-Stop Scaffolding; p.35 (left) Genie Industries, (right) Scott Schneider; p.37 (bottom) Streimer Sheet Metal Works, Inc.; p.39 (bottom) Hilti Corporation; p.41 (top) Midstate Education and Service Foundation, (bottom) Tape Tech Tools; p.43 (both photos) Midstate Education and Service Foundation; p.49 (bottom) Expanded Shale, Clay, and Slate Institute; p.51 (top) Messer Construction, (bottom) Spec Mix Inc.; p.53 (top) Scott Fulmer, (middle/bottom) Jennifer Hess; p.55 (top) Wood's Powr-Grip; p.59 Cal/OSHA; p.61 (all photos) Cal/OSHA; p.63 (all photos) Midstate Education and Service Foundation; p.65 (bottom) Quickpoint, Inc.; p.67 (bottom) ErgoAir, Inc.; p.69 (top) Messer Construction; p.71 (middle/bottom) Midwest Tool and Cutlery Co.; p.73 (bottom) Slip-On Lock Nut Co. and Morton Machine Works.

Tip Sheet Contributors

Tip Sheet #1. Jim Albers, MPH, CIH, NIOSH, Cincinnati, OH, and Cherie Estill, MS, PE, NIOSH, Cincinnati, OH.

Tip Sheet #2. Scott Schneider, MS, CIH, Laborers' Health and Safety Fund of North America, Washington, DC, and Jim Albers, MPH, CIH, NIOSH, Cincinnati, OH.

Tip Sheet #3. Jennifer Hess, DC, PhD, University of Oregon Labor Education and Research Center, Eugene, OR, and Jim Albers, MPH, CIH, NIOSH, Cincinnati, OH.

Tip Sheet #4. Kate Stewart, MS, and Steve Russell, MS, Seattle, WA, and Build It Smart, Olympia, WA.

Tip Sheet #5. Peter Vi, MS, Construction Safety Association of Ontario, Etobicoke, Ontario, Canada, and Jim Albers, MPH, CIH, NIOSH, Cincinnati, OH.

Tip Sheet #6. Phil Lemons and Kelly True, Streimer Sheet Metal, Portland, OR, and Jim Albers, MPH, CIH, NIOSH, Cincinnati, OH.

Tip Sheet #7. Charles P. Austin, MS, CIH, Sheet Metal Occupational Health Institute Trust (SMOHIT), Alexandria, VA.

Tip Sheet #8. Greg Shaw, Midstate Education and Service Foundation, Ithaca, NY.

Tip Sheet #9. Greg Shaw, Midstate Education and Service Foundation, Ithaca, NY.

Tip Sheet #10. Dan Anton, PhD, PT, ATC, University of Iowa, College of Public Health, Department of Occupational and Environmental Health, Iowa City, IA.

Tip Sheet #11. Pamela Entzel, JD, MPH, Center to Protect Workers' Rights, Silver Spring, MD, Jim Albers, MPH, CIH, NIOSH, Cincinnati, OH, and Cherie Estill, MS, PE, NIOSH, Cincinnati, OH.

Tip Sheet #12. Jennifer Hess, DC, PhD, University of Oregon Labor Education and Research Center, Eugene, OR, and the Center to Protect Workers' Rights, Silver Spring, MD.

Tip Sheet #13. Jim Albers, MPH, CIH, NIOSH, Cincinnati, OH, and Cherie Estill, MS, PE, NIOSH, Cincinnati, OH.

Tip Sheet #14. Adapted from the booklet *Easy Ergonomics: A Guide to Selecting Non-Powered Hand Tools* (2004), a joint publication of the California Dept. of Occupational Safety and Health (Cal/OSHA) and NIOSH. Cincinnati, OH: U.S. Department of Health and Human Services, Centers for Disease Control and Prevention, National Institute for Occupational Safety and Health, DHHS (NIOSH) Publication No.2004-164.

Tip Sheet #15. Greg Shaw, Midstate Education and Service Foundation, Ithaca, NY.

Tip Sheet #16. Jim Albers, MPH, CIH, NIOSH, Cincinnati, OH, and Cherie Estill, MS, PE, NIOSH, Cincinnati, OH.

Tip Sheet #17. Jim Albers, MPH, CIH, NIOSH, Cincinnati, OH, and Cherie Estill, MS, PE, NIOSH, Cincinnati, OH.

Tip Sheet #18. Jim Albers, MPH, CIH, NIOSH, Cincinnati, OH, and Cherie Estill, MS, PE, NIOSH, Cincinnati, OH.

Tip Sheet #19. Charles P. Austin, MS, Sheet Metal Occupational Health Institute Trust (SMOHIT), Alexandria, VA, Jim Albers, MPH, CIH, NIOSH, Cincinnati, OH, and Cherie Estill, MS, PE, NIOSH, Cincinnati, OH.

Tip Sheet #20. Jim Albers, MPH, CIH, NIOSH, Cincinnati, OH, and Cherie Estill, MS, PE, NIOSH, Cincinnati, OH.

Reviewers

NIOSH wishes to acknowledge the following early reviewers of this document. Reviewers' organizations are listed for identification only. While their suggestions have improved the quality of the material, the authors accept full responsibility for the content: Tom Alexander (Independent Electrical Contractors, National Safety Committee), Tony Barsotti, CSP (Temp-Control Mechanical Corporation), Bruce Bowman, PE (Independent Electrical Contractors, National Safety Committee), Stephen Hecker, PhD (University of Washington-Seattle), Ira Janowitz, MS, CPE (Lawrence Livermore National Laboratory), Rashod Johnson, PE (Masonry Contractors Association of America), Phil Lemons, CSP (Streimer Sheet Metal), John Masarick (Independent Electrical Contractors), Mike McCullion, CSP (Sheet Metal and Air Conditioning Contractors National Association), Jim McGlothlin, PhD, CPE (Purdue University), Gary Mirka, PhD (Iowa State University), Brian L. Roberts, CSP, CIE (Independent Electrical Contractors), Kristy Schultz, MS, CIE (California State Compensation Insurance Fund).

Table of Contents

Foreword

Construction is a physically demanding occupation, but a vital part of our nation and the U.S. economy. In 2006, the total annual average number of workers employed in construction rose to an all-time high of nearly 7.7 million, according to U.S. Bureau of Labor Statistics data. This large workforce handled tasks that range from carrying heavy loads to performing repetitive tasks, placing them at risk of serious injury. The physically demanding nature of this work helps to explain why injuries, such as strains, sprains, and work-related musculoskeletal disorders, are so prevalent and are the most common injury resulting in days away from work.

Although the construction industry presents many workplace hazards, there are contractors in the U.S. who are successfully implementing safety and health programs to address these issues, including work-related musculoskeletal disorders.

The safety and health of all workers is a top priority for NIOSH. This booklet is intended to aid in the prevention of common job injuries that can occur in the construction industry.

The solutions in this booklet are practical ideas to help reduce the risk of repetitive stress injury in common construction tasks. While some solutions may need the involvement of the building owner or general contractor, there are also many ideas that individual workers and supervisors can adopt.

There are sections on floor and ground-level work, overhead work, material handling, and hand-intensive work. For each type of work, "simple solutions" for various tasks are described in a series of "Tip Sheets." The solutions consist mostly of materials or equipment that can be used to do the job in an easier way. Each Tip Sheet describes a problem, one possible solution, its benefits to the worker and employer, how much it costs, and where it can be purchased. All these solutions are readily available and are actually in use today in the U.S. construction industry.

We encourage both contractors and workers to consider the "simple solutions" in this booklet and look for ways you can adapt them to your own job and worksite.

John Howard, M.D.
Director
National Institute for Occupational Safety and Health
Centers for Disease Control and Prevention

Why This Booklet?

This booklet is intended for construction workers, unions, supervisors, contractors, safety specialists, human resources managers—anyone with an interest in safe construction sites. Some of the most common injuries in construction are the result of job demands that push the human body beyond its natural limits. Workers who must often lift, stoop, kneel, twist, grip, stretch, reach overhead, or work in other awkward positions to do a job are at risk of developing a *work-related musculoskeletal disorder (WMSD)*. These can include back problems, carpal tunnel syndrome, tendinitis, rotator cuff tears, sprains, and strains.

To aid in the prevention of these injuries, this booklet suggests many simple and inexpensive ways to make construction tasks easier, more comfortable, and better suited to the needs of the human body.

Example of a "simple solution." This ironworker uses a tool that automatically ties rebar with the pull of a trigger. The extended handle lets him work while standing upright. No leaning, kneeling, stooping, or hand twisting are necessary.

Did You Know . . . ?

- Construction is one of the most hazardous industries in the United States.

- The number of back injuries in U.S. construction was 50% higher than the average for all other U.S. industries in 1999 (CPWR, 2002).

- Backaches and pain in the shoulders, neck, arms, and hands were the most common symptoms reported by construction workers in one study (Cook et al, 1996).

- Material handling incidents account for 32% of workers' compensation claims in construction, and 25% of the cost of all claims. The average cost per claim is $9,240 (CNA, 2000).

- Musculoskeletal injuries can cause temporary or even permanent disability, which can affect the worker's earnings and the contractor's profits.

The "Tip Sheets" in this booklet show how using different tools or equipment may reduce the risk of injury. All of the items described in this booklet have been used on working construction sites. Given the nature of construction, some solutions here may not be appropriate for all worksites. Sometimes solutions discovered for one trade can be modified for other trades.

This booklet provides general information regarding the methods some construction contractors have used to reduce workers' exposures to risk factors for work-related musculoskeletal disorders. The examples described in this booklet may not be appropriate for all types of construction work. The use of the tools and equipment described in the booklet does not ensure that a musculoskeletal disorder will not occur. The information contained in this booklet does not produce new obligations or establish any specific standards or guidelines.

Our goal has been to describe solutions that are also cost-effective. Although the cost of some of the solutions here exceeds $1,000, which may be too high for some contractors, we believe successful implementation will lead to a quick recovery of the investment in many cases.

Oh, My Aching Body!

Construction work is hard work, and construction workers feel the results. In one survey, seven out of ten construction workers from 13 trades reported back pain, and nearly a third went to the doctor for it (Cook et al, 1996).

Back pain, carpal tunnel syndrome, tendinitis, rotator cuff syndrome, sprains, and strains are types of musculoskeletal disorders. *Work-related musculoskeletal disorders (WMSDs)* are caused by job activities and conditions, like lifting, repetitive motions, and work in confined areas. All of these are part of construction work. WMSDs can become long-term, disabling health problems that keep you from working and enjoying life. Not only do these injuries hurt your body, but they can reduce your earnings and your employer's profit.

You have an increased risk of these injuries if you often:

- Carry heavy loads

- Work on your knees

- Twist your hands or wrists

- Stretch to work overhead

- Use certain types of tools

- Use vibrating tools or equipment.

On top of that, tight deadlines mean a fast pace. Pushing the pace increases your risk even more.

A study of workers' compensation claims filed in Washington State between 1990–98 reported that the highest risks for developing a WMSD were "in industries characterized by manual handling and forceful repetitive exertions." According to the study, construction work accounted for 10 of the top 25 sectors in need of interventions to prevent neck, back, and upper extremity WMSDs (Silverstein, 1998).

One insurance company reported that 29% of insured mechanical and electrical contractors' workers' compensation claims were due to WMSDs. A quarter of those claims resulted in temporary or permanent disability. The insurer also reported that WMSD claims for electrical contractors average around $6,600 for each WMSD, while the average claim for a mechanical contractor was around $7,300 (NIOSH 2006).

Many people in construction believe that sprains and strains are just the nature of the business. But new tools and materials are now available that can make work less risky and increase productivity. This booklet shows some of the solutions, large and small, to WMSDs.

As you read this booklet, the solutions may or may not apply to your specific jobsite or trade. You will need to review cost, quality, and site-specific information to make sure that the solution will meet your needs. Also, these ideas can be adapted. Notice the principles involved: What kinds of activities are most likely to cause injuries? How can they be minimized?

Sometimes a small change in tools, equipment, or materials can make a big difference in preventing injuries. We wish you the best as you strive to make improvements to the work you do and your worksite.

> *NIOSH believes that better work practices and tools can reduce the frequency and seriousness of sprains and strains among construction workers.*
>
> *These suggestions can be adapted for your own jobsite.*
>
> **SAFER • HEALTHIER • PEOPLE™**

What Is Ergonomics?

The goal of the science of ergonomics is to find a "best fit" between the worker and job conditions. Ergonomics tries to come up with solutions to make sure workers stay safe, comfortable, and productive. These usually involve changing tools, equipment, materials, work methods, or the workplace itself. Ergonomics is a new topic for the construction industry, but the ideas have been around for many years. For example, in 1894 the split-level scaffold was designed for masonry work in the U.S. to reduce workers' frequent bending. This new scaffold system was designed to improve workers' productivity by reducing the time spent in awkward positions. There is still a strong case for using ergonomic improvements both to reduce workers' exposure to risk factors for WMSDs and to improve their productivity.

Ergonomics looks at how:

Physical abilities of the human body	A R E R E L A T E D T O	Work tasks
— and —		Tools, equipment, and materials
Limitations of the human body		The job environment

Work-Related Musculoskeletal Disorders

Work-related musculoskeletal disorders (WMSDs) are the leading cause of disability for people in their working years. They can be caused by frequently working in a way that puts stress on the body, such as:

- Gripping
- Kneeling
- Lifting
- Working in awkward positions
- Applying force
- Repeating movements
- Bending
- Working overhead
- Twisting
- Using vibrating equipment.
- Squatting
- Over-reaching.

The best way to reduce WMSDs is to use the principles of ergonomics to redesign tools, equipment, materials, or work processes.

Simple changes can make a big difference. Using ergonomic ideas to improve tools, equipment, and jobs reduces workers' contact with those factors that can result in injury. When ergonomic changes are introduced into the workplace or job site, they should always be accompanied by worker training on how to use the new methods and equipment, and how to work safely.

Do You Need an Ergonomics Program?

Many ergonomics experts recommend that employers and joint labor-management groups develop their own ergonomics programs to analyze risk factors at the worksite and find solutions. These programs may operate as part of the site's health and safety program, or may be separate. An ergonomics program can be a valuable way to reduce injuries, improve worker morale, and lower workers' compensation costs. Often, these programs can also increase productivity.

There may be a particularly urgent need for an ergonomics program at your site if:

- Injury records or workers' compensation claims show excessive hand, arm, and shoulder problems; low back pain; or carpal tunnel syndrome.

- Workers often say that some tasks are causing aches, pains, or soreness, especially if these symptoms do not go away after a night's rest.

- There are jobs on the site that require forceful actions, movements that are repeated over and over, heavy lifting, overhead lifting, use of vibrating equipment, or awkward positions such as raising arms, bending over, or kneeling.

- Other businesses similar to yours have high rates of work-related musculoskeletal disorders.

- Trade magazines or insurance publications in your industry frequently cover these disorders.

Effective ergonomics programs have included the following elements:

- Employer commitment of time, personnel, and resources

- Someone in charge of the program who is authorized to make decisions and institute change

- Active employee involvement in identifying problems and finding solutions

- A clearly defined administrative structure (such as a committee)

- A system to identify and analyze risk factors

- A system to research, obtain, and implement solutions such as new equipment

- Worker and management training

- Medical care for injured workers

- Maintaining good injury records

- Regular evaluation of the program's effectiveness.

Education and training programs have been developed for construction general contractors by the Associated General Contractors, the United Brotherhood of Carpenters and Joiners, the Sheet Metal Occupational Health Institute, and the Laborers' Union. Although the problems and solutions described in these organizations' materials may be specific to a sector or trade, you may find them useful when developing your own ergonomics program.

For additional information on developing an ergonomics program, see *Elements of Ergonomics Programs* (NIOSH Pub. No. 97-117) at *www.cdc.gov/niosh/docs/97-117*.

Simple Solutions for Floor and Ground-Level Work

The Problem

On some construction jobs you need to work close to the ground or floor. For example, you may have to stoop or kneel when installing or finishing slabs, decks, or floor coverings.

Bending, stooping, kneeling, or squatting can cause pain in your lower back or knees. Over time you may develop a serious muscle or joint injury. Your risk is higher if you stoop or kneel often or for long periods of time. It is also higher if you twist your body while working in these positions.

These positions can also make it harder to do your job. When stooping or kneeling, you can't lift, push, or pull as much weight without putting stress on your body.

Injuries & Disorders

Below are some of the injuries you may develop when you work at floor level.

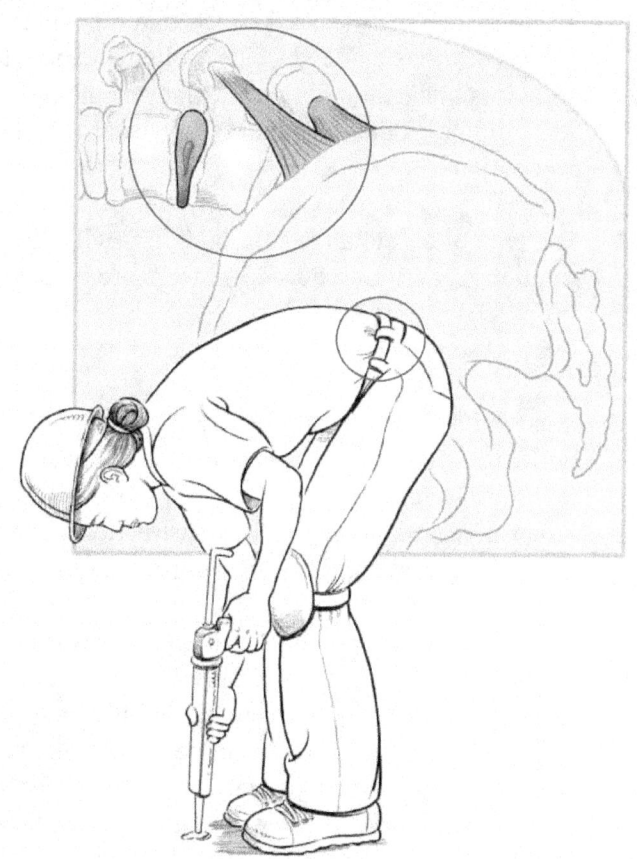

Lower back. Your spine runs from the top of your neck down to your lower back. It is made up of many bones called *vertebrae,* one below another. Between the vertebrae are *joints* and *discs*. These give your back flexibility so it can move. The discs are flexible because they have a substance like jelly inside.

When you bend forward, your back muscles work harder and the *ligaments* (long fibers supporting the back muscles) flex and stretch. The discs get squeezed. As they are squeezed, they can press on different parts of the spine, including nerves. This can cause back pain. If you bend forward over and over for months or years, the discs are weakened, which may lead to disc rupture (or "herniation").

Twisting your body while bending puts even more pressure on the discs, and more stress on the cartilage and ligaments, especially when you are exerting force to lift, push, or pull objects.

Knee. The muscles in your knee are connected to your leg by *tendons*. Between the tendons and bones are small sacs of fluid called *bursa*. They lubricate the knee so it moves easily.

Continual stress on your knee can cause the bursa to get squeezed, swollen, stiff, and inflamed (*bursitis*). This stress can also cause the knee tendons to become inflamed, resulting in pain (*tendinitis*).

Tasks that involve frequent stooping, kneeling, or squatting increase your risk of developing bursitis, tendinitis, or arthritis in the knee. The risk of arthritis increases for workers who already have had a knee injury and work in these positions.

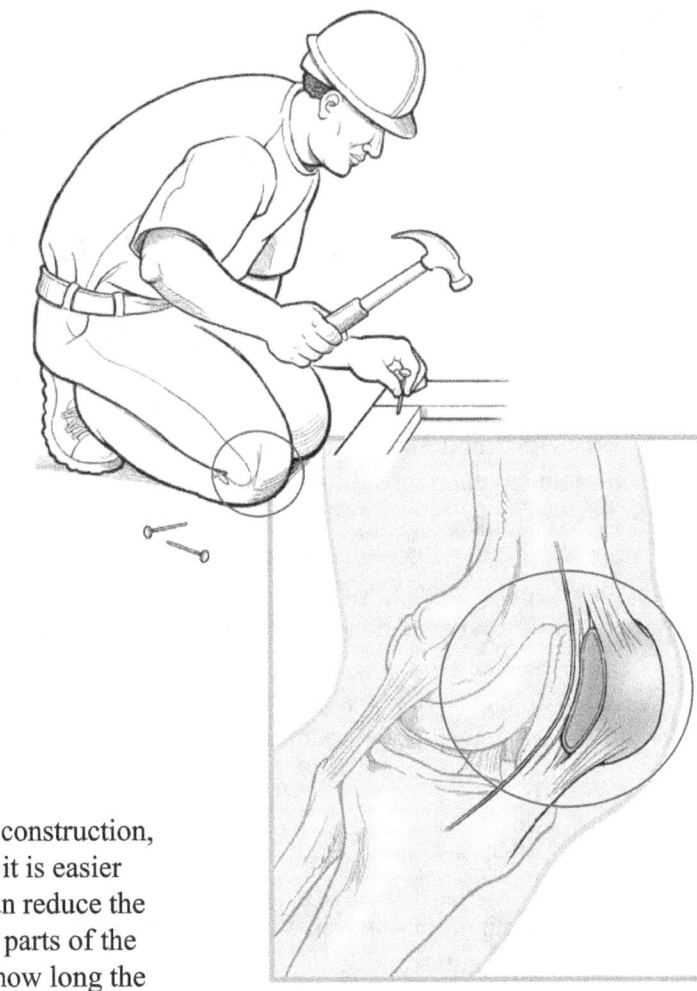

Some Solutions

Floor-level work cannot be eliminated from construction, but it is possible to change how you do it so it is easier on your body. Solutions are available that can reduce the level of stress on your back, knee, and other parts of the body. They may also reduce how often and how long the body is subjected to this stress. Many of the solutions can also eliminate other potential safety hazards and increase productivity.

The type of task and the site conditions will determine which solutions are best for you. A few possible solutions for specific floor-level tasks are explained in Tip Sheets #1–5.

General solutions for doing floor-level work with less risk of injury include:

Change materials or work processes. One of the most effective solutions may be to use materials, building components, or work methods that are less labor-intensive, so the task takes less time and you therefore kneel and stoop for a shorter period. Because there may be cost, contract, and engineering issues involved, an individual construction worker or subcontractor usually cannot make a decision like this. Changes may require the approval of the architect, engineer, building owner, or general contractor.

However, individual workers often *can* change the way they do the work. Sometimes people work on the floor because it is the only large flat work area available. The floor is used as a workbench for assembling, mixing, or other tasks. This is common, for example, when assembling sheet metal ducts or building rebar cages. This increases the amount of stooping or kneeling that is necessary. Rather than stooping to the floor, try to raise the work to waist height using tables, sawhorses, or other equipment. It is possible to make your own improvised workbench out of materials you have available.

Change tools and/or equipment. For example, use tools with extension handles that let you stand up while doing a floor-level task. In a few cases, cost and site conditions may restrict the use of such tools.

Change work rules and provide training. Contractors can set site rules that require the use of benches, tables, or sawhorses to raise the work up so less kneeling and stooping are necessary. Rules can also require that materials be stored off the ground. Limits can be placed on the total time that workers do floor-level work without a break. In cases where kneeling on a hard surface cannot be avoided, knee pads or some other type of padding should be used. Also, a policy of providing ergonomics training may help workers more quickly identify potential problems and find effective solutions.

Example: Gurney converted to work table

Fastening Tools that Reduce Stooping

Problem: Stooping to use screw gun

The Problem

When working at floor or ground level, construction workers often use screw guns and other fastening tools that require stooping, bending, kneeling, or squatting for long periods of time. Working repeatedly in these positions can result in fatigue, pain, and injury.

Your lower back and knees are the areas at greatest risk of a muscle or joint injury when you stoop, bend, kneel, or squat for prolonged periods. Your risk is increased if you have to lift, push, or pull while stooping.

One Solution

Use an **auto-feed screw gun with an extension** that allows you to stand upright while working. Standing while you work keeps your spine and knees in a neutral position, minimizing strain and muscle fatigue. Many stand-up tools have adjustable lengths to fit workers of different heights. Stand-up screw guns that automatically feed the screws are available. Powder-actuated fastening tools (PATs) can be used with a stand-up handle provided by the manufacturer.

How It Works

A **screw gun with an extension** can be used to secure subflooring, false floors, and decking; to construct concrete forms; and to do other wood-to-wood jobs. You can also

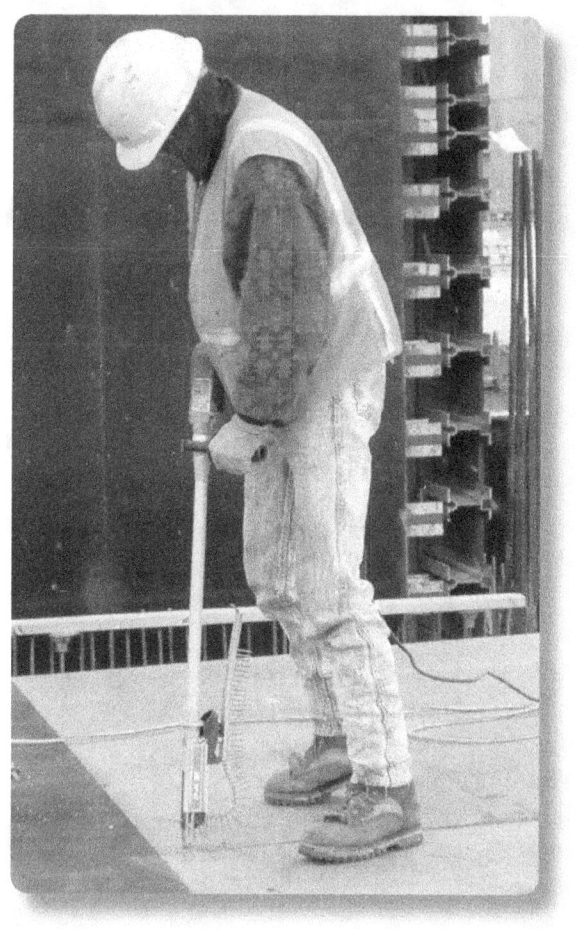
Solution: Autofeed stand-up screw gun

use it for drywall and some metal-to-metal work. Screws for these guns come on easy-loading, collated strips that are self-feeding. They load in seconds without requiring you to bend down, and the newer models have been improved so they do not jam like older models. There are models with fixed and telescoping extensions. Some use extensions that can be removed, allowing the screw gun to be used by itself for walls.

A **powder-actuated fastening tool** with a stand-up handle can be used to fasten metal track to concrete decks for interior steel framing, to install plywood onto concrete as a substrate for wood floors, to attach lumber to concrete and masonry, and to make steel-to-steel connections. These are a fast, reliable, efficient fastening method that can be used independent of weather conditions. They fire a .27-caliber explosive charge to drive their fasteners. The fasteners are made from hardened steel and have a knurled shaft to anchor them securely in the base material. Pre-drilling holes is not necessary. The driving depth can be adjusted for varying jobsite conditions. Hearing protection is advised when firing PATs.

Benefits for the Worker and Employer

Workers who spend less time in a stooped position, or kneeling, have less chance of developing lower back and knee injuries. Productivity is also improved. Studies have shown that autofeed stand-up screw guns are about twice as fast at placing screws as traditional screw guns. Both stand-up screw guns and PAT tools with stand-up handles have improved since they were first introduced and are now more dependable and easier to use. Screws are more expensive than nails and may not be cost-effective for some jobs. However, using screws may improve the quality of construction on other jobs, such as installing subfloors.

Approximate Cost

Stand-up screw guns are $200–400. PAT fastening tools with stand-up handles are $500–700. The PAT handles can also be purchased separately

For More Information

- Products related to this solution are described at *www.cpwr.com/simple.html*. Products also may be found on the internet using the following search terms:

    Stand-Up Screw Guns: "screw gun extension"

    Powder-Actuated Tools with Stand-Up Handles: (tool manufacturer) + "stand-up handle"

- Local contractor tool and equipment suppliers or rental companies may be another source of information on products.

- For general information on this solution, check *www.cpwrconstructionsolutions.org* and *www.elcosh.org*.

Motorized Concrete Screeds

The Problem

When you hand screed concrete, you work bent over, and you have to use a strong grip to pull the board over the wet concrete. Your arms and shoulders exert a lot of force over and over.

Doing this work often or for a long period of time increases your chance of fatigue and pain. It puts major stresses on your back, knees, hands, arms, and shoulders, which may lead to serious muscle or joint injuries.

Problem: Hand screeding

One Solution

Use a **motorized screed** (also called a **vibratory screed**). You can work standing upright, and operating the screed takes much less effort than hand screeding.

This type of screed eliminates both screeding in a stooped position and the need for repeated arm and shoulder movements.

Solution: Motorized screeding

How It Works

The motorized screed consists of a blade or plow that floats on the concrete, one or two gasoline motors that vibrate the blade, metal support tubing, and handles to hold when you operate it.

It works best for small to medium-sized jobs.

Benefits for the Worker and Employer

A motorized screed should reduce a worker's chance of developing muscle and joint injuries. This equipment greatly reduces the physical effort needed for hand screeding, and eliminates the frequent and prolonged stooping. Little effort is needed to move the plow over the concrete surface.

Screeding with motorized equipment can be faster than hand screeding. Many contractors report improvements in productivity. The vibration of the blade improves the consolidation of the concrete and reduces time spent "bull floating" the surface.

There are some drawbacks. Although you can work around electrical or plumbing stubs, some hand screeding may still be necessary. Also, it can be difficult to move the screed to and from the work location. A single-engine screed weighs around 50 lbs., and can be awkward to lift and carry. Some screeds have a quick-release system to remove the plow from the frame, which makes carrying easier.

Vibration can also be a problem. It is important to protect workers from hand-arm vibration syndrome (HAVS), a nerve disorder that can become disabling. NIOSH measured vibration levels on three types of motorized screeds. Two had the gasoline engine placed at the bottom of the frame and above the plow. One screed had the engine placed on a single shaft, and the operator held the shaft below the engine. Vibration levels for the two types with the engine at the bottom were below the current recommended guidelines to prevent HAVS. The third screed, which was also older and poorly maintained, gave off much higher vibration that could exceed current HAVS guidelines. Higher vibration levels are expected when the engine is connected to the frame or shaft that the operator must grip. When buying a motorized screed, ask about vibration levels and test drive the screed.

Approximate Cost

A single-engine motorized screed costs around $1,500. A twin-engine model costs around $4,000 and requires two operators.

For More Information

- Products related to this solution are described at *www.cpwr.com/simple.html*. Products also may be found on the internet using the following search terms: "power screed," "vibratory screed," or "concrete screed."

- Local contractor tool and equipment suppliers or rental companies may be another source of information on products.

- For general information on this solution, check *www.cpwrconstructionsolutions.org* and *www.elcosh.org*.

Rebar-Tying Tools

The Problem

Ironworkers tie rebar by hand with pliers and tie wire. This work requires repeated, fast hand and arm movements while applying a lot of force. If you tie rebar at ground level, you also have to work in a stooped position, with your body bent deeply forward.

Tying rebar by hand increases your chance of developing hand-wrist disorders due to the high hand forces used to grip pliers, the rapid hand movements used to wrap and twist wire, and the high pressure on the hand and fingers when twisting and cutting wire. If you work at ground level, you also are at risk of low back injuries from frequent and prolonged stooping and bending.

Problem: Tying rebar by hand

One Solution

Use a **rebar-tying tool**. This lowers your risk of hand and wrist injury because it eliminates the frequent rapid hand motions required when using pliers. Some rebar tiers allow you to work standing up, so there is less stress on your low back due to stooping and bending.

Solution: Rebar-tying tool with extension handle

How It Works

Both manual and battery-powered rebar-tying tools are currently available.

Battery-powered rebar tiers automatically fasten the bars together with tie wire. They can be used whenever a simple "wrap and twist" tie is needed. However, they do not provide the strength of "saddle" or "figure 8" ties.

Several companies offer power rebar tiers. With one tool design, you press the trigger and the tool feeds wire around the bars and then twists and cuts the wire. These models are not stand-up tools, but an adjustable extension handle is available.

A second tool is a stand-up power tier that uses coiled spring wire to hold the bars together. The tool automatically "screws" (or spins) flat coiled wire around the intersecting bars. This tool was designed using ergonomic principles.

Benefits for the Worker and Employer

Workers should experience fewer injuries. Studies conducted by NIOSH and the Construction Safety Association of Ontario (Canada) compared manual methods and one model of power tying tool, and showed that using the power tool may reduce the risk of injury to workers' hands, wrists, and low back.

There have been documented increases in productivity. The NIOSH-Ontario studies found that power tying tools can tie rebar twice as fast as hand tying. Actual productivity increases will depend on the type of work and the frequency of tying. Also, contractors and rod busters who used the model of power tool involved in the studies reported they preferred it to manual tying for flat work. Before using one of these tying tools, make sure the ties are approved for the job you will be doing.

Approximate Cost

Wire feeding tiers are under $2,700 and wire costs around 2 cents per tie. Tiers using coiled spring wire are under $1,300 and wire costs around 3 cents per tie. Powered models generally require extra batteries and chargers, which may be included in the price.

For More Information

- Products related to this solution are described at *www.cpwr.com/simple.html*. Products also may be found on the internet using the following search terms: "rebar tying system" or "rebar tier."

- Local contractor tool and equipment suppliers or rental companies may be another source of information on products.

- For general information on this solution, check *www.cpwrconstructionsolutions.org* and *www.elcosh.org*.

Kneeling Creepers

The Problem

Many construction tasks require frequent kneeling, squatting, or stooping because the work is close to the floor. Kneeling on a hard surface puts a lot of direct pressure on your knee, while squatting puts stress on the tendons, ligaments, and cartilage of the knee joint. Working in either position often or for long periods of time can lead to knee problems, including knee osteoarthritis.

If you work in a stooped position, there is stress on your lower back as well as your knees, possibly leading to back pain and even a serious back injury.

Problem: Kneeling to work near floor

One Solution

Use a portable **kneeling creeper** with chest support. When the job requires kneeling or squatting to work at floor level, these devices will reduce the stress to your knees, ankles, and lower back.

Solution: Laying tile with kneeling creeper

How It Works

Kneeling creepers are available with removable seats and cushioned knee supports. They are very low and have 2 to 3 inch casters. The knee supports on one model are only ¾ inch above the floor. The cushioned knee supports reduce the pressure on your knees, just as ordinary knee pads do.

Some models are available with an adjustable cushioned chest support. It is useful when doing prolonged floor-level jobs like tile setting and concrete patching. It helps support your weight, reducing back strain and some of the pressure on your knees.

Benefits for the Worker and Employer

Kneeling creepers provide support when work must be done in awkward and stressful positions. They reduce stress on the knees and lower back, and can help prevent serious muscle and joint problems. Since work can be done with less discomfort and pain, productivity often also increases.

Kneeling creepers allow workers to move around more easily and quickly, and may also have an area where tools can be conveniently placed.

These devices can be used to assist injured workers retuning to the job, since they can work with less stress to their knees and back.

Approximate Cost

Kneeling creepers without the chest support cost around $200 and the optional adjustable chest support is around $75.

For More Information

- Products related to this solution are described at *www.cpwr.com/simple.html*. Products also may be found on the internet using the following search terms: "kneeling creeper."

- Local contractor tool and equipment suppliers or rental companies may be another source of information on products.

- For general information on this solution, check *www.cpwrconstructionsolutions.org* and *www.elcosh.org*.

TIP SHEET #5

Adjustable Scaffolding for Masonry Work

The Problem

Masons often need to stoop to pick up brick, block, and mortar and place them on a wall. This work can require a lot of bending and twisting of the body.

You have to bend deeper and twist your body more often if you keep materials below hip height, or lay brick or block on a section of wall below hip height.

Frequent stooping causes fatigue and puts stress on your lower back. This stress increases your chance of developing low back pain or serious back injury. Your risk of injury is even higher if you also twist your body quickly, especially when holding heavy objects.

Problem: Conventional unguarded frame scaffolding

One Solution

Use **split-level adjustable scaffolding**. This allows a brick or block mason to stoop less because the materials and work surface are both kept near waist height, which is more comfortable and stresses your body less. Split-level adjustable scaffolds are available for jobs ranging from small single-story residential work to large high-rise building projects. This equipment may not be appropriate for all jobs.

Solution: Masons finish top course on split-level adjustable scaffolding

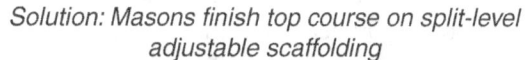

How It Works

Adjustable scaffolding has separate platforms for the worker and the materials. Since the worker platform can be raised or lowered, the materials and the work surface can both be positioned around the height of the worker's waist. Small scaffolds can be raised manually using hand jacks or a hand winch. Taller scaffolds can be raised using a powered winch.

Benefits for the Worker and Employer

Masons reduce their chance of developing low back pain or back injury. Workers spend less time handling materials because brick, block, and mortar are moved shorter distances. Workers expend less physical effort and report less fatigue at the end of the day. They also appreciate the wider platforms, which give more room to move around.

Mason tenders who build conventional frame scaffolds also get important benefits. Split-level adjustable scaffolds reduce the heavy physical labor involved in continuously changing the height of a frame scaffold. There is less lifting and carrying of frames and heavy boards, and less moving materials to the ground and back. Split-level adjustable scaffolds also reduce the danger of falling that mason tenders face when building frame scaffolds.

Measurable increases in productivity have been reported when split-level adjustable scaffolds are used. One study found that masons' productivity increased 20% when using the scaffolds, and some contractors have reported even greater increases in productivity.

Approximate Cost

Costs vary depending upon how the work platform is raised (manual vs. powered), the height of the masonry wall, and the overall size of the project. Heavy-duty scaffolding for three-story commercial work runs about $300 per lineal foot. Light-duty scaffolding for residential and light commercial work of the same height runs about $200 per lineal foot. Contractors who have used these systems say that productivity increases allow them to recover the cost, but actual benefits will vary.

For More Information

- Products related to this solution are described at *www.cpwr.com/simple.html*. Products also may be found on the internet using the following search terms: "adjustable scaffolding."

- Local contractor tool and equipment suppliers or rental companies may be another source of information on products.

- For general information on this solution, check *www.cpwrconstructionsolutions.org* and *www.elcosh.org*.

Simple Solutions for Overhead Work

The Problem

On some construction jobs you need to work overhead, reaching up with one or both arms raised above your shoulders. Your head may be tilted back, looking up to see what you are doing. Whether you are drilling, driving fasteners, or finishing drywall, overhead work puts stress on your shoulders and neck. Eventually it may lead to serious muscle and joint injuries.

You are at risk of injury if you do this work often or for long periods of time. The risk is increased if you frequently hold tools, equipment, or materials above the height of your shoulders, or if you twist your body while your arms are in an awkward, raised position.

When you work with your arms raised, injuries are even more likely if you have to use repetitive movements or a lot of force. For example, using some hand tools overhead requires you to repeat the same motions over and over, as well as apply force, while you are reaching up. Lifting, holding, and positioning heavy or awkward objects while your arm is raised can also require force.

Working overhead also may reduce your ability to do the job safely and productively. For example, you are at risk of many types of injury if your vision is obstructed, if you have an unstable footing, or if you have trouble holding or positioning a tool.

Injuries & Disorders

Below are some of the injuries you may develop when you work overhead.

Shoulder. Shoulder pains and injuries are usually the result of overworking the shoulder. When you keep your arm raised above your shoulder (or keep your arm stretched out), your shoulder begins to ache after a short time. It tires easily.

The muscles in your shoulder are connected to your arm by *tendons*. Between the tendons and bones are small sacs of fluid called *bursa*. They lubricate the shoulder so it moves easily. Continual stress on your shoulder can cause the bursa to get squeezed, swollen, stiff, and inflamed (*bursitis*). Bursitis

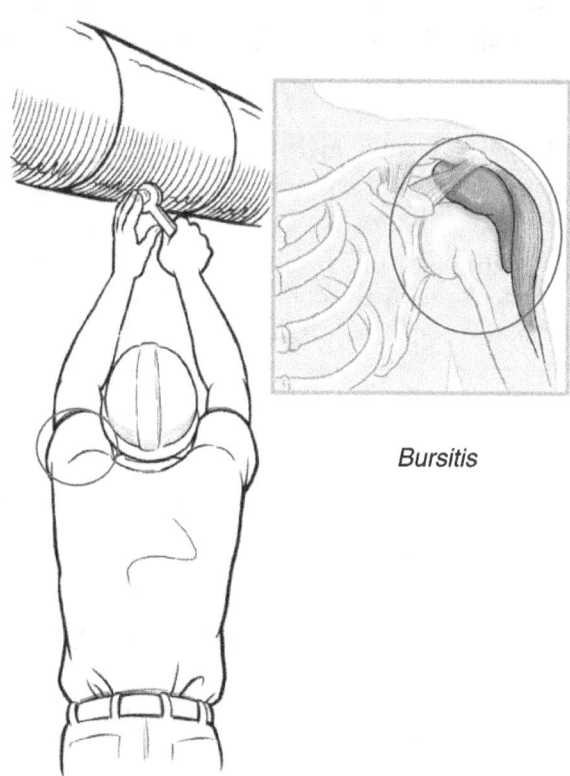

Bursitis

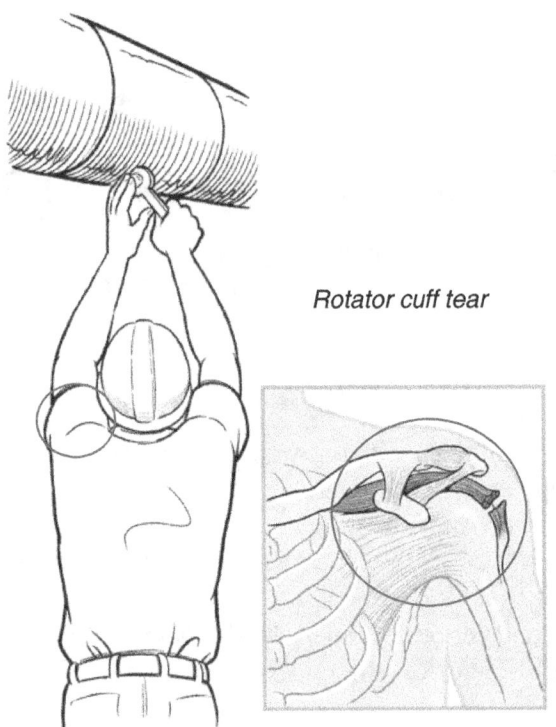

Rotator cuff tear

can make it painful, or even impossible, to raise your arm. Continual stress on the shoulder can also cause the shoulder tendons to become inflamed, resulting in pain (*tendinitis*).

Another shoulder injury is the *rotator cuff tear*. The rotator cuff is a group of four muscles and their tendons that wrap around the front, back, and top of the shoulder joint. These let the shoulder function through a wide range of motions. Stress on the shoulder may cause them to tear, which can make routine activities difficult and painful.

A NIOSH review of studies found that the risk of developing shoulder pain or a shoulder muscle or joint disorder is increased by the combination of frequently working with raised shoulders (60° or more), using repetitive arm or shoulder movements while in this position, and applying force while in this position.

Neck. The neck is a complicated structure composed of seven bones called *cervical vertebrae,* one below another. It also has *cartilage*, *nerves*, *muscles*, and *ligaments* (long fibers supporting the muscles). When you keep your neck bent forward or backward, or bend it frequently, the muscles work harder and the ligaments flex and stretch. Eventually the ligaments can partially tear, resulting in a *neck sprain*.

Another common condition is *tension neck syndrome*. This is muscle strain that results from long periods of looking up. It can cause neck stiffness, muscle spasms, and pain in the neck or radiating from the neck.

It is also possible to develop *arthritis* in the neck. The risk of arthritis increases for workers who already have had a neck injury and still do overhead work.

According to a 1997 NIOSH study, your risk of developing neck pain or a neck musculoskeletal disorder is increased by the combination of frequently working with your neck flexed (15° or more), using repetitive movements, and applying force while in this position.

Some Solutions

Overhead work cannot be eliminated from construction, but it is possible to change how you do it so it is easier on your body. Solutions are available that can reduce the level of stress on your shoulders, neck, and arms. They may also reduce how often and how long the body is subjected to this stress. Many of the solutions can also eliminate other potential safety hazards and increase productivity.

The type of task and the site conditions will determine which solutions are best for you. A few possible solutions for specific overhead tasks are explained in Tip Sheets #6–9.

General solutions for doing overhead work with less risk of injury include:

Change materials or work processes. One of the most effective solutions may be to use materials, building components, or work methods that are less labor-intensive, so the task takes less time and you reach overhead for a shorter period. For example, installing embedded concrete inserts into ceiling forms would eliminate the prolonged overhead drilling needed to place all-thread rods for ceiling systems. An individual construction worker or subcontractor usually cannot make a decision like this. Certain changes may require the approval of the building owner, architect, engineer, or general contractor.

Change tools and/or equipment. For example, use bit extensions for drills and screw guns that allow you to hold the tool at waist or shoulder level rather than above your head. Use mechanical lifts or hoists to raise and position building materials rather than lifting them manually. Or use a lift to raise yourself so you are closer to the work. In a few cases, cost and site conditions may restrict the use of such equipment.

Change work rules and provide training. Contractors can encourage the use of equipment like extensions, lifts, and hoists that reduce the need for workers to raise their arms. Site rules can limit the amount of time that workers do overhead work without a break. Also, a policy of providing ergonomics training may help workers more quickly identify potential problems and find effective solutions.

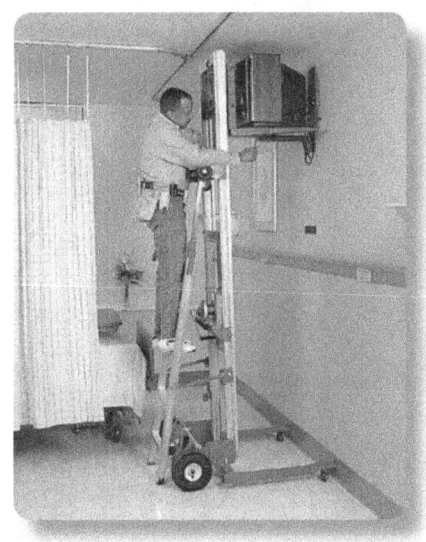

Mechanical lifts reduce the strain of holding and positioning objects

Person-lifts provide stable platforms for overhead work and eliminate handling of ladders and scaffolding

TIP SHEET #6

Bit Extension Shafts
for Drills and Screw Guns

The Problem

If you use a drill or screw gun for overhead work, you are forced to keep your arms and neck in fixed, awkward, hard-to-hold positions. You have to push upward with a heavy tool above your shoulders, using your shoulder muscles instead of your biceps.

This work can put stress on your arms, neck, shoulders, and back. It can lead to fatigue and serious muscle or joint injuries.

One Solution

Use a **bit extension shaft** for the drill or screw gun, so you can hold the tool below your shoulder and closer to your waist.

You strain your arms, neck, shoulders, and back less because you don't have to hold the tool above your shoulders or work in an awkward position. You work with your upper arms held close to your sides, and your hands in front of your body. You are pushing with your biceps muscles instead of your shoulders.

How It Works

The extension is like a normal socket, but longer. It can be made from carbon steel tube stock. One end of the tube is welded to a piece that fits into the drill or screw gun chuck. An impact socket is welded onto the other end of the tube. A bit can then be

Problem: Drilling overhead

Solution: Using extension

inserted into the socket. PVC pipe is used as a sleeve over the extension. Part of the PVC pipe slides over the bit. The sleeve protects your hands from the spinning shaft and gives you better control of the tool. You hold the tool with one hand and the sleeve with the other.

Extensions are available from several manufacturers in many different lengths and diameters. Lengths generally run from 12 to 40 inches. It is also possible to make your own extension. Before using an extension for drilling or fastening, you should determine whether it is appropriate for the job.

Benefits for the Worker and Employer

Bit extensions should reduce a worker's chance of developing muscle and joint injuries. Workers benefit from not having to hold a heavy tool above their shoulders for long periods. The extension itself weighs less than two pounds. Using the extension is easy after a little practice. One company that uses the extensions says that their workers like the way the extension prevents shoulder strain.

However, an extension does not solve the problems caused by holding your neck tilted back to look up at your work. To cut down on neck strain, avoid looking up when you don't have to. To be safe, make sure you push the extension onto the drill chuck tightly, before you turn on the power. In a screw gun, the screw must also be tight in the bit.

Approximate Cost

You can make your own bit extension, which will cost $1–2 for materials plus labor. Manufactured bit extensions cost from approximately $12 for a 12-inch model to $45 for a 24-inch model.

For More Information

- Products related to this solution are described at *www.cpwr.com/simple.html*. Products also may be found on the internet using the following search terms: "bit extension shaft."

- Local contractor tool and equipment suppliers or rental companies may be another source of information on products.

- For general information on this solution, check *www.cpwrconstructionsolutions.org* and *www.elcosh.org*.

Extension Poles for Powder-Actuated Tools

The Problem

Using powder-actuated fastening tools (PATs) for overhead work may lead to serious shoulder, arm, and hand injuries.

You work with your arms above your shoulders, an awkward position that may cause sore muscles and joints. Sometimes you have to hold this difficult position for a long time, or repeat the position over and over during your shift. This can cause fatigue and eventually lead to arm and shoulder problems like bursitis or rotator cuff tendinitis.

The recoil shock transmitted from the tool to your hand, arm, and shoulder put you at even greater risk of injury. Also, when you shoot fasteners at ceiling height you must work from a ladder, scaffold, or power lift, which have their own hazards.

One Solution

Use an **extension pole** for overhead work. This is a fixed height or modular pole attached to the powder-actuated tool. Using the extension, the tool is entirely out of your hand. All you have to do is squeeze the trigger.

You no longer need to raise your arms above your shoulders and hold them there to work on the ceiling. The extension does it for you.

The extension lets you keep a more neutral body posture. Your arms are closer to your

Problem: Using PAT overhead

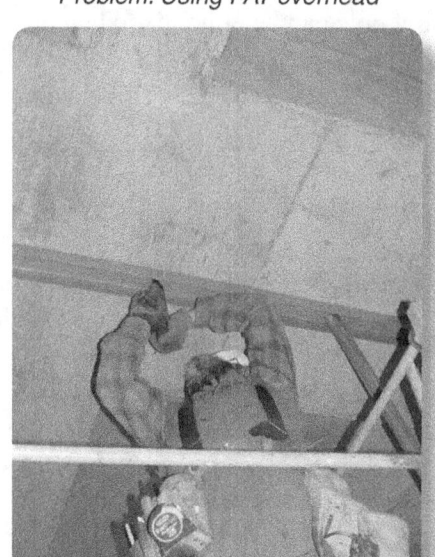

Solution: PAT with modular extension

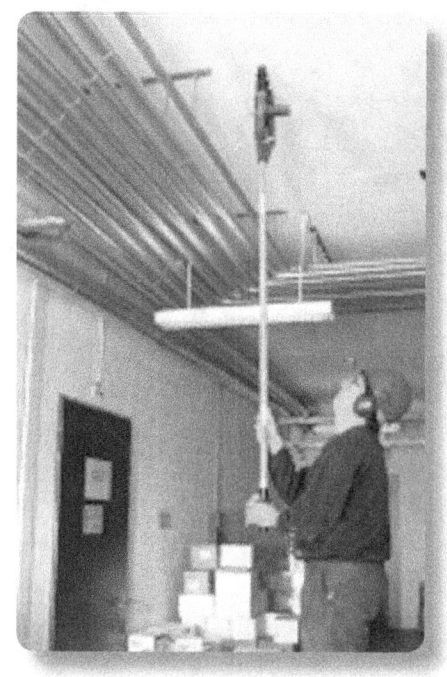

body and below your shoulders. This cuts down on the risk of injury to your shoulders, arms, and hands. There is also less recoil shock directed to your shoulders and neck. In addition, you can work on the ground rather than using a ladder, scaffold, or lift.

How It Works

The powder-actuated tool attaches to the top of the pole. A hand trigger (like a motorcycle brake) is attached to the bottom of the pole. You need only squeeze the trigger to operate the tool.

The extension pole can be either fixed height or adjustable. Lengths of available extensions range from 3 to 18 feet, though poles longer than 8 feet may be too difficult to position and control. The gun must be held tight to the substrate at a 90-degree angle until the firing is complete.

Benefits for the Worker and Employer

Workers have less chance of developing a shoulder, arm, and hand injury. With the tool attached to the pole, you can keep your arms below your shoulders. You will also feel less recoil shock at your shoulder. Moreover, with the gun further from your head, the noise exposure will be lower. Using the pole will also keep your eyes and face further from the concrete dust and debris. You do still need to look up to position the shot, which puts some strain on your neck.

Productivity may improve when there is less setup time using the tool extension, because there are no are no ladders, scaffolds, or lifts needed.

Approximate Cost

A modular pole assembly costs from $300–400. However, if ladders, scaffolds, or lifts were rented for this work in the past, you will no longer have this cost.

For More Information

- Products related to this solution are described at *www.cpwr.com/simple.html*. Products also may be found on the internet using the following search terms: (PAT manufacturer) + "pole tool."

- Local contractor tool and equipment suppliers or rental companies may be another source of information on products.

- For general information on this solution, check *www.cpwrconstructionsolutions.org* and *www.elcosh.org*.

Spring-Assisted Drywall Finishing Tools

Simple Solutions

The Problem

If you use typical flat and corner "mudboxes" for drywall finishing, you need to use a lot of strength to push the compound out of the box. The forceful, repetitive pushing motions combined with reaching overhead can cause fatigue. Eventually this work can lead to serious wrist, arm, shoulder, and back injuries.

With a flat box, you often have to push hard while bending your wrist and back. When you bend like this while pushing forcefully, you increase your chance of developing a muscle or joint injury. Your risk is higher when you do the same job over and over.

Finishers say that corner boxes require them to push even harder than flat boxes, because you have to squeeze the compound into a tighter space.

Problem: Hand drywall finishing with mudbox

One Solution

Use a **spring-assisted finishing tool**. This does most of the pushing for you. Spring-assisted flat boxes and corner tools are available that cut down significantly on the strain caused by pushing.

With spring-assisted flat boxes, the springs provide up to 75% of the force needed to push the compound onto the wall. Spring-assisted corner tools provide 100% of the force needed to finish corners.

Solution: Spring-assisted mudbox

Overhead Work

How It Works

Spring-assisted flat boxes look and work just like regular boxes. The box wheels activate the power assist feature. Springs on the outside of the box pull on levers. The levers then push on a pressure plate, and the compound squirts out when the wheels touch the wall. The boxes come in several different widths and with different handle lengths.

Spring-assisted corner finishers have hydraulically damped springs. When you turn a handle, the spring pushes the compound through a normal corner-finishing head. All you need to do is guide the tool down the corner.

You use a regular pump to fill both types of spring-assisted tools with compound.

Benefits for the Worker and Employer

A spring-assisted mudbox should reduce a worker's chance of developing a muscle or joint injury. More than 80% of the workers involved in one study liked using the new tools much better than the older ones. After using both new tools, all workers in the study said they were not as tired as when using the regular tools. Most said they had less pain.

The same study suggested that there were gains in productivity. Spring-assisted flat boxes gave the same results and were just as fast as regular boxes. However, they led to less worker fatigue and pain, and workers said they got more done because they were less tired. They also were able to use the new boxes for a longer time without becoming fatigued. In addition, most workers said that the new spring-assisted corner tool was faster and easier to use, compared to the older corner boxes.

Approximate Cost

You can rent or purchase these tools. Rentals cost about the same as for regular boxes. One manufacturer sells a set of three spring-assisted flat boxes for about $1,300 and a spring-assisted corner finisher for about $1,400.

For More Information

- Products related to this solution are described at *www.cpwr.com/simple.html*. Products also may be found on the internet using the following search terms: "drywall tool" + "spring assisted."

- Local contractor tool and equipment suppliers or rental companies may be another source of information on products.

- For general information on this solution, check *www.cpwrconstructionsolutions.org* and *www.elcosh.org*.

Pneumatic Drywall Finishing Systems

The Problem

Finishing drywall by hand requires lots of effort and repetition. Your body is forced into awkward positions that can lead to serious wrist, shoulder, arm, and back injuries.

Working with bent wrists, and with your back bent or twisted, is common in hand finishing. You repeat certain difficult hand, arm, and back movements over and over. Some tools used in this work (such as flat and corner "mudboxes") may also cause problems because you use a lot of strength to push on them.

The combination of pushing and working in an awkward position causes fatigue as well as tired and sore muscles. Eventually it can increase your chance of developing a muscle or joint injury.

Problem: Hand drywall finishing with mudbox

One Solution

Use a **pneumatic drywall finishing system**. You can avoid hand finishing, and you won't have to use flat and corner boxes. Although the work will still require some awkward positions and motions, you will not be exerting high forces at the same time or for as long a time. An air compressor gives enough pressure to force the drywall compound through the flat- and corner-finishing heads of the pneumatic system.

The pneumatic system requires much less physical effort than hand finishing, and it cuts down on the repeated wrist, arm, and

Solution: Pneumatic finishing system

and back movements. It also takes less effort to use than the automatic tools. You do not need to push with the same strong force used with boxes. You only have to guide the powered tool down the corner.

How It Works

An air compressor forces the drywall compound from the hopper into a tube. You do not need to hand pump the compound. Air forces the compound through the head of the tool when you pull the trigger. The heads give the same quality of finish that boxes do. The tool requires an airline from the hopper to the applicator. You also need electrical power for the air compressor.

Benefits for the Worker and Employer

These systems should reduce a worker's chance of developing muscle and joint injuries. In one study, most workers who tried the pneumatic system liked it better than the regular tools. Workers reported that their muscles were not as tired, and they did not have as much pain after using the pneumatic system.

There can also be an improvement in productivity. Workers in the study said the pneumatic system is faster than the older finishing boxes. Some workers said that it took a while to learn the new system. Although using the system was slow at first, it became faster within two months.

There are some drawbacks to the pneumatic system. The air and electrical lines may restrict your movement while you work.

The new system also takes longer to move, set up, and clean than hand finishing equipment. Therefore, it may not be as practical for small finishing jobs.

Approximate Cost

Pneumatic finishers start around $3,500 and cost more depending on features.

For More Information

- Products related to this solution are described at *www.cpwr.com/simple.html*. Products also may be found on the internet using the following search terms: "drywall tool" + "pneumatic."

- Local contractor tool and equipment suppliers or rental companies may be another source of information on products.

- For general information on this solution, check *www.cpwrconstructionsolutions.org* and *www.elcosh.org*.

Simple Solutions for Lifting, Holding, and Handling Materials

The Problem

On many construction sites, workers spend time lifting, carrying, holding, pushing, or pulling loads of material. Although it is common today to use mechanical devices for some of this work, a lot of materials are still handled manually. Sometimes it is not possible to use mechanical material handling devices due to site conditions.

If you lift and carry materials often or for long periods of time, there is constant stress on your back and shoulders. Eventually you may develop a serious muscle or joint injury. You are at risk if you often handle materials that are heavy and/or bulky, carry materials long distances, stoop downward to pick up heavy objects, or stretch upward while holding them. Your risk is higher if you twist your body when handling heavy items.

You may also develop an injury if you frequently push or pull heavy carts, dollies, or other transport equipment.

Injuries & Disorders

Below are some of the injuries you may develop when you do manual material handling.

Back. Low back pain, and more serious musculoskeletal injuries to the back, can occur suddenly or develop over a period of time. For example, sudden quick movements, especially while handling heavy objects, may lead immediately to painful muscle strains. These strains may develop into serious injuries when the muscles are not allowed to heal and are exposed to additional stress.

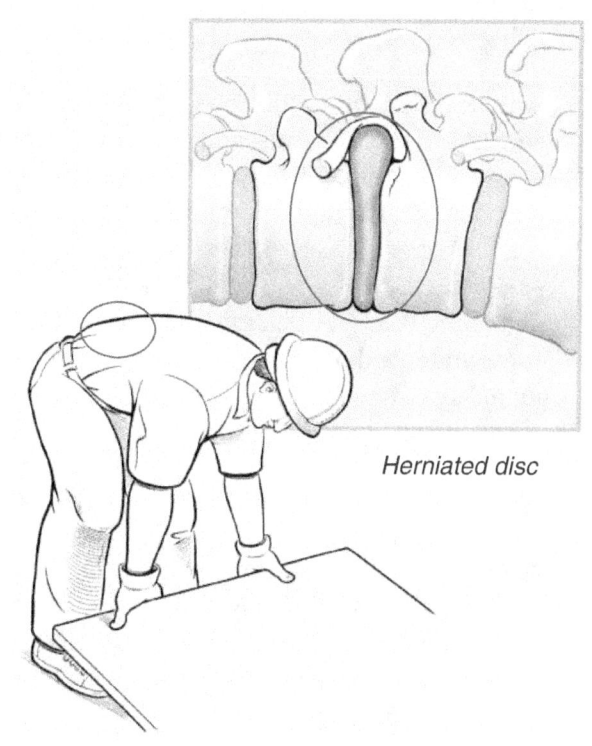

Herniated disc

Your spine runs from the top of your neck down to your lower back. It is made up of many bones called *vertebrae,* one below another. Between the vertebrae are *joints* and *discs.* These give your back flexibility so it can move. The discs are flexible because they have a substance like jelly inside.

When you lift, bend forward, stretch upward, or stretch outward, your back muscles work harder and the *ligaments* (long fibers supporting the back muscles) flex and stretch. The discs get squeezed. As they are squeezed, they can press on different parts of the spine, including nerves. This can cause back pain. If you bend forward over and over for months or years, the discs are weakened, which may lead to disc rupture (or "herniation").

Twisting your body while bending puts even more pressure on the discs, especially when you are exerting force to lift, push, or pull objects.

Shoulder and neck. Carrying even light loads above your shoulders may quickly lead to tired and sore neck and shoulder muscles. The risk of developing a more serious neck or shoulder problem increases when you do this work frequently or for long periods of time. Carrying or resting heavy objects on your shoulders may stress the shoulder and neck muscles and cause injury where the load contacts your body.

Trapezius muscle

The muscles in your shoulder are connected to your arm by *tendons*. Between the tendons and bones are small sacs of fluid called *bursa*. They lubricate the shoulder so it moves easily. Continual stress on your shoulder can cause the bursa to get squeezed, swollen, stiff, and inflamed (*bursitis*). Bursitis can make it painful, or even impossible, to raise your arm.

Continual stress on the shoulder can also cause the shoulder tendons to become inflamed, resulting in pain (*tendinitis*).

Another common condition is *tension neck syndrome*. This is a type of muscle strain that can cause neck stiffness, muscle spasms, and pain in the neck or radiating from the neck. It affects the *trapezius* muscle, a large, thin muscle that runs from the upper back through the shoulder area to the neck. You may notice a tender "knot" in this muscle as well as stiffness and pain.

Arms, hands, and wrists. If you carry heavy objects with hard sharp edges, they can dig into your skin and injure the soft tissues in your hands. Or, if you carry objects that are hard to grip and hold, they may force your hand or wrist into awkward, stressful positions and cause disorders like tendinitis or carpal tunnel syndrome.

Some Solutions

Manual material handling is still common in construction, but it is possible to change how you do it so it is easier on your body. Solutions are available that can reduce the level of stress on your back, shoulder, neck, and other parts of your body. They may also reduce how often and how long your body is subjected to this stress. Many of the solutions can also eliminate other potential safety hazards and increase productivity.

The type of task and the site conditions will determine which solutions are best for you. A few possible solutions for specific material handling tasks are explained in Tip Sheets #10–13.

General solutions for doing material handling with less risk of injury include:

Change materials or work processes. One of the most effective solutions may be to use materials, building components, or work methods that are less labor-intensive. There are alternative materials that can be handled without requiring a lot of physical strength, an awkward posture, or repetitive motion. For example, half-weight bags of Portland cement and lightweight concrete masonry blocks are currently available in many areas. An individual construction worker or subcontractor usually cannot make a decision to switch materials. Certain changes may require the approval of the building owner, architect, engineer, or general contractor.

Change tools and/or equipment. You can buy or rent material handling devices for all aspects of construction. Devices include special round handles and cushioned grips for carrying heavy objects; powered and non-powered carts and dollies for indoor or outdoor use; rolling carts to move sheet materials, pipes, or conduit; and stands and jacks to hold materials during installation.

Mechanical, hydraulic, and vacuum lifts are available in a variety of sizes and styles. Some allow relatively easy positioning of components and materials.

Ergonomic Guidelines for Manual Material Handling (DHHS/NIOSH Publication No. 2007-131) describes many different types of material handling and transport equipment. This booklet can be accessed at *www.cdc.gov/niosh/docs/2007-131/pdfs/2007-131.pdf*.

In a few cases, cost and site conditions may restrict the use of such equipment.

Power vacuum lifter avoids manual lifting

Change work rules. For example, contractors can require that materials be stored at a convenient height off the ground and transported in most situations with mechanical devices. Improved planning of laydown areas and materials storage can minimize the number of times materials need to be moved.

Provide training and related programs. A policy of providing ergonomics training may help workers more quickly identify potential problems and find effective solutions.

Workplace exercise programs are popular in the construction industry. Although they may be a part of any effort to prevent muscle and joint disorders, exercise programs are not a substitute for other solutions. No studies have shown that they prevent injuries by themselves. Studies indicate only that exercise may have a short-term effect on reducing low back pain. There also is no evidence supporting the use of "body mechanics education" as an effective means to prevent back pain or serious back disorders. In edition, NIOSH does not recommend the use of back belts to prevent back injuries.

Training in the NIOSH lifting guidelines is especially important. NIOSH recommends that one person lift no more than 51 lbs. when the lifting can be done using the following "best practices":

- When you pick up or set down a load, don't reach more than 10 inches away from your body.

- Don't twist your body.

- Lift with your legs, not your back. Keep your back as straight as possible.

- Lift the load using a solid two-handed grip.

When lifting, holding, and positioning materials on a construction site you can't always follow these "best practices." In that case, the 51 lb. weight limit needs to be lowered. See the "Applications Manual for the Revised NIOSH Lifting Equation" (1997) for more information on how to use the guidelines. This information should be passed along to workers in training programs.

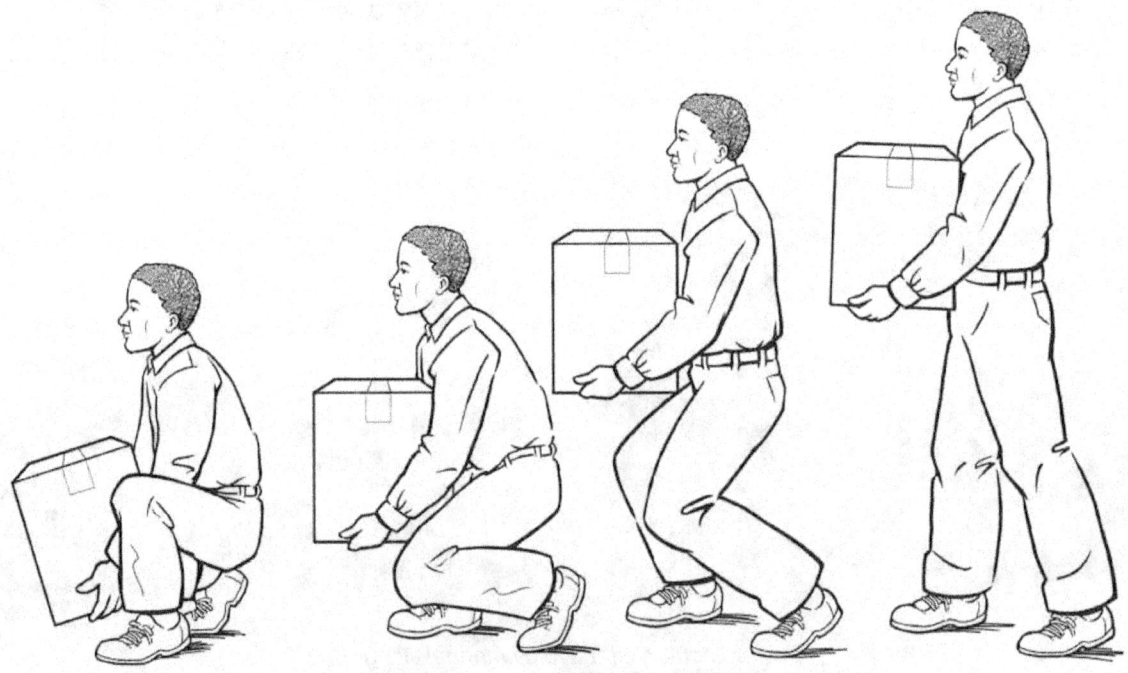

Lightweight Concrete Block

The Problem

A regular concrete block (also called a *concrete masonry unit*, or CMU) can weigh up to 50 pounds, depending on size. For masons and mason tenders, lifting and placing CMUs can cause fatigue and put strain on the low back, hands, and arms. If you do this work often, you may be at risk of a serious muscle or joint injury.

The risk depends on how many units you handle, how heavy they are, how often you work with them, how low they are stored, and how high you have to reach to place them on the course. You have even more risk if you twist your body when lifting or holding CMUs, or if you lift or hold them with one hand.

Problem: Laying standard concrete block

One Solution

Use **lightweight concrete block**. Units weigh 30-40% less than regular block without sacrificing strength or performance. Working with lightweight block can improve your output during the day and still decrease the total weight you lift. Less weight means you will be less tired and there will be less stress on your back, hands, and arms.

Solution: Types of lightweight block

How It Works

The aggregate used for lightweight block is made from shale, clay, and/or slate. These materials are expanded in a rotary kiln at temperatures over 1000° C.

The block is structurally strong, stable, and durable, yet also light in weight and a good insulator. The block density is only 40–50 pounds per cubic foot. An ordinary block made from rock and sand has a density of 105–115 pounds per cubic foot. Lightweight block meets or exceeds the specifications required of regular heavy concrete block (American Society for Testing and Materials (ASTM) C 90 Standards Specifications for Load-bearing Concrete Masonry Units).

Benefits for the Worker and Employer

Laying lightweight CMUs reduces a worker's fatigue and lowers stress on the back and arm muscles. One study looked at how concrete block of different weights affects muscle stress. Masons built two walls. One used lightweight CMUs and the other regular CMUs. When workers built the lightweight CMU wall, they had less back and arm muscle stress. The difference was greatest when lifting the block to the top of high walls.

There can also be a gain in productivity. According to the National Concrete Masonry Association (NCMA), "lighter weight units resulted in higher productivity rates (other factors being equal)."

Approximate Cost

Lightweight block costs slightly more per unit than standard block. However, since masons and mason tenders can work faster and better, there should be a reduction in labor cost. This can account for up to 80% of the finished wall cost. Shipping and handling costs may be lower as well.

For More Information

- Products related to this solution are described at *www.cpwr.com/simple.html*. Products also may be found on the internet using the following search terms: lightweight "concrete (or) masonry (or) block."

- Local contractor tool and equipment suppliers or rental companies may be another source of information on products.

- For general information on this solution, check *www.cpwrconstructionsolutions.org* and *www.elcosh.org*. The Expanded Shale, Clay, and Slate Institute (ESCSI) in Salt Lake City, UT also has more information at *www.escsi.org*.

Pre-Blended Mortar and Grout Bulk Delivery Systems

The Problem

When laborers mix mortar or grout in the traditional way, they have to lift heavy cement bags and shovel sand into the mixer. They usually repeat these motions over and over. Full cement bags weigh around 100 pounds, and workers may handle more than 100 bags a day.

Workers are at risk of back pain, shoulder pain, and even disabling muscle or joint injuries. The injuries can be the result of one-time accidents, but they usually develop over a long period of time.

Your risk of injury from lifting and shoveling depends on the weight of the load, the number of loads you lift, how long you do this work, and where the bags are placed. If you have to stoop down to a low pallet to pick up bags, or reach above your shoulders to dump them, your risk is especially high. You have even more risk if you twist your body while lifting.

One Solution

Use **pre-blended mortar and grout mix** for the job. This can be delivered to the site in bulk and doesn't require lifting bags or shoveling sand.

Bulk pre-blended mortar and grout can be used with conventional mixers or with European-style silo mixers. All dry ingredients are handled mechanically, with either a forklift or boom truck, eliminating the risk of injury due to manual handling.

Problem: Traditional method of loading mixer

Solution: Loading a silo system

How It Works

Pre-blended dry ingredients (including sand, pigments, and admixtures) are delivered to the site in 2,000 – 3,000 pound bulk bags. These are moved by forklift or boom truck over a funnel-shaped silo that straddles a conventional mortar mixer. The mix is released into the silo by pulling a hitch pin on the bag. To begin mixing, the mixer operator simply pulls a handle that opens the silo's discharge slide gate. The pre-blended material is gravity-fed from the silo directly into the mixer below. No electricity is needed. Only water needs to be added to produce the mortar and grout.

Benefits for the Worker and Employer

Laborers have less risk of disabling injuries due to constant manual lifting. Productivity is also increased because time-consuming manual handling of bags and sand is eliminated. One laborer may be able to tend two or three mixers.

With silo systems, bags don't break or leak. The product is more consistent because it is pre-mixed. There is no problem of sand freezing in winter. Silo dispensers can save space on congested jobsites and reduce material theft. Most silo systems are easily moved around the jobsite. Disposal of bags is not a problem since all bulk bags are removed by the supplier and either reused or recycled.

These systems do not eliminate the risk of silica dust exposure, but dust curtains are available to reduce the release of silica-containing dust. Specific safety procedures must be followed when loading the silo, climbing the silo's ladder, and moving the system.

Approximate Cost

Contractors estimate that using these systems adds about 7 - 8% to mortar cost. However, the additional cost may be offset by improvements in efficiency and productivity gains. Pre-blended mortar systems may not be cost-effective for smaller jobs. A supplier, however, should be able to help you determine if this product is appropriate for your work.

For More Information

- Products related to this solution are described at *www.cpwr.com/simple.html*. Products also may be found on the internet using the following search terms: (silo or bulk) "delivery systems."

- Local contractor tool and equipment suppliers or rental companies may be another source of information on products.

- For general information on this solution, check *www.cpwrconstructionsolutions.org* and *www.elcosh.org*.

Skid Plates to Move Concrete-Filled Hoses

Simple Solutions

The Problem

Charged concrete hoses are heavy and pulling them takes a lot of strength. The latches on a hose may snag on rebar. Workers must sometimes bend down and lift the hose to free it.

Pulling, lifting, and moving sections of hose can force your body into awkward positions and put strain on your lower back and knees. If you have to use jerking motions or twist your body while doing this work, there is even more strain on your back. Handling concrete hoses, especially for long periods of time, may cause fatigue, back pain, and even serious muscle or joint injuries.

One Solution

Skid plates (also known as "hose placing discs") may be useful when concrete boom pumps and other alternative ways of moving the concrete cannot be used. Skid plates are two-foot diameter concave metal disks that are placed under the hose couplings. They have a cradle to hold the hose and handles for carrying. They decrease the friction with the rebar matting underneath and make the hose easier to pull. They also prevent the latches on the hose from catching on the rebar.

How It Works

Laborers usually move concrete-filled hoses across rebar matting by pulling on ropes attached to the hose, or by using long metal hooked rods.

Problem: Pulling concrete hose without skid plate

Solution: Pulling hose with skid plate and hook

Skid plate

Lifting, Holding, & Handling Materials

Skid plates slide more easily across the rebar matting, reducing the friction. Pulling is easier. Also, hose couplings do not catch on the rebar matting. This decreases the need for laborers to jerk the hose or bend over to free it.

From four to six skid plates should be used near the pour end of the hose. They are most effective when the hose is secured to each plate. You can fasten the hose to a plate with rebar tying wire or rubber bungee cords. Both can quickly be removed when necessary. Using *unsecured* skid plates may lead to more bending, awkward positions, and back strain.

Benefits for the Worker and Employer

At least one study has found that using skid plates secured to the hose can reduce stress to the low back that otherwise would increase the chance of developing a serious injury.

Use of secured skid plates does not result in loss of productivity. It takes only moments to place skid plates under hoses and secure them. If workers are less fatigued from pulling heavy hoses, productivity may actually increase.

There are a few drawbacks. For example, it is still possible for skid plates to catch on Nelson studs (4" tall steel rods welded to the subflooring to reinforce the concrete). Skid plates reduce the physical stress of pulling a hose, but they don't get rid of it. The plates should be used only when the charged hose cannot be moved with a boom, crane, or motorized concrete placement equipment.

Approximate Cost

Prices run about $200–300 per plate.

For More Information

- Products related to this solution are described at *www.cpwr.com/simple.html*. Products also may be found on the internet using the following search terms: "concrete" + "hose placing disc."

- Local contractor tool and equipment suppliers or rental companies may be another source of information on products.

- For general information on this solution, check *www.cpwrconstructionsolutions.org* and *www.elcosh.org*.

Vacuum Lifters for Windows and Sheet Materials

The Problem

Manually installing large windows and sheet materials requires workers to handle heavy and bulky objects. You may need to lift and carry them some distance to the installation site. When installing them, you may have to use a lot of strength to hold them while they are placed and secured.

This work puts stress on your back and shoulders, which can lead to serious muscle and joint injuries. Injuries can be even more serious when you have to work in awkward positions or hold materials for a long period of time. Manually placing windows and sheet materials may also lead to hand injuries.

Problem: Installing window using manual suction cups still requires lifting

One Solution

Use **vacuum lifters** to install windows and other flat panels. Vacuum lifters eliminate the need to manually lift and position heavy and awkward materials.

A vacuum lifter can be attached to a forklift, or to a small counter-balanced crane built in the shop. It can also be attached to a larger crane for outdoor work.

How It Works

There are both non-powered and powered vacuum lifter systems available with load capacities from 375   1400 pounds. Non-powered "hand cup" systems lift and carry the load using manually-operated, pump-

Solution: Using powered vacuum lift

style vacuum cups attached to a specially-designed frame. Some of these frames permit loads to be rotated and tilted. In some cases, the cups can be removed from the frame so they can be used individually for lifting and carrying.

Although some contractors use manual systems, a more typical system is a powered lifter using cups with a cord-free 12-volt vacuum pump. "Below-the-hook" powered vacuum lifters are a little more expensive but feature rotation and tilt.

Benefits for the Worker and Employer

Large window units and other panels can be installed without the usual physical stress that comes with lifting, carrying, holding, and positioning heavy objects. Using lifters will help reduce the possibility that a worker will develop a muscle or joint injury.

A vacuum lifter also keeps the fingers and arms from being caught in pinch points when positioning and setting the window or panel.

There should be a gain in productivity, since workers will be less fatigued and able to install more windows or panels. There may also be less damage to windows and other materials.

Approximate Cost

Manually-operated "hand cup" frames featuring rotation and tilt cost around $1,200. You will spend about $300 total for four 9-inch pump-style vacuum cups to use with them.

A basic four-cup DC-powered lifter costs around $2,500. Below-the-hook vacuum lifters are available with various options and configurations, ranging from $3,000 – 7,000.

For More Information

- Products related to this solution are described at *www.cpwr.com/simple.html*. Products also may be found on the internet using the following search terms: "vacuum lifters."

- Local contractor tool and equipment suppliers or rental companies may be another source of information on products.

- For general information on this solution, check *www.cpwrconstructionsolutions.org* and *www.elcosh.org*.

Simple Solutions for Hand-Intensive Work

The Problem

Construction workers usually spend a lot of time gripping tools or materials with one or both hands. This work can put stress on your hand, wrist, and/or elbow, causing discomfort and pain. Eventually you may develop a serious muscle or joint injury. Your ability to use your hands and wrists may be reduced, and you may even become permanently disabled.

You are at risk of injury if you often use a forceful grip on tools, bend your wrist when using them, or move your wrist rapidly or repetitively. Injuries also can result if you frequently hold vibrating tools, or if tool handles that are hard or sharp often press into your hand, wrist, or arm.

Gripping tools and other materials can be physically demanding, repetitive work. It may injure the muscles, tendons, and cartilage of your hand, wrist, and elbow. Damage to the nerves and blood vessels can also occur.

If you experience soreness or pain, and continue doing the work without allowing your muscles and tendons to rest and heal, the pain may get worse and you may eventually develop a serious disorder.

Injuries & Disorders

Below are some of the injuries you may develop when you do hand-intensive work.

Tendinitis. Most of the muscles that move your hand and fingers are actually in your forearm. These muscles are connected to the hand and fingers by tendons, which are like cords passing through your wrist.

You can strain the tendons in your wrist if you frequently exert strong force with your hand, bend your wrist while working, or repeat the same wrist movements over and over. If this strain continues over time, you may develop tendinitis. Tendinitis makes it painful to use your hand, especially to grasp things.

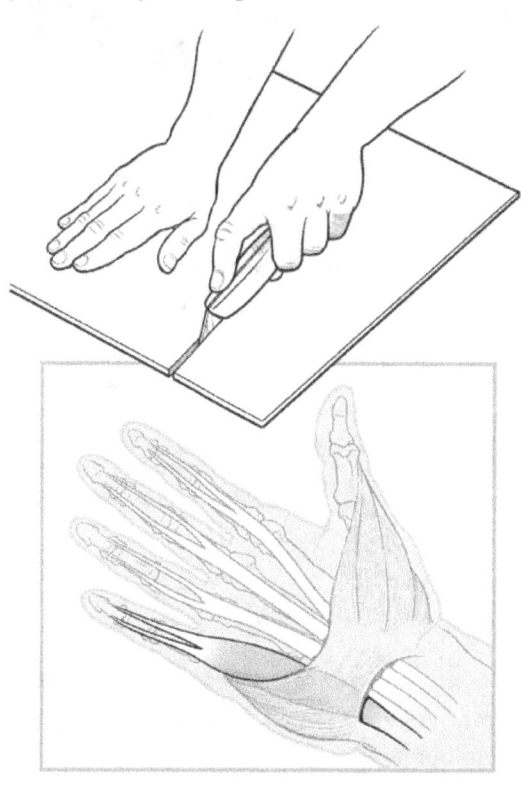

Carpal tunnel syndrome. The carpal tunnel is an area in your wrist that is surrounded by bone and tissue. A nerve and several tendons pass through this tunnel. If you have tendinitis and your tendons swell, there is less room in the tunnel for the nerve. When the nerve is squeezed this way, the condition is called carpal tunnel syndrome. It often leads to pain, tingling, or numbness in your hand, wrist, or arm. These symptoms are often felt at night.

If left untreated, carpal tunnel syndrome can weaken the hand and make it very difficult to grasp things or even use that hand at all.

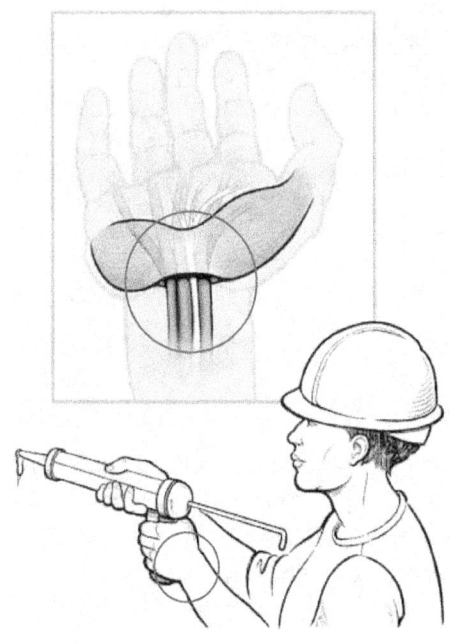

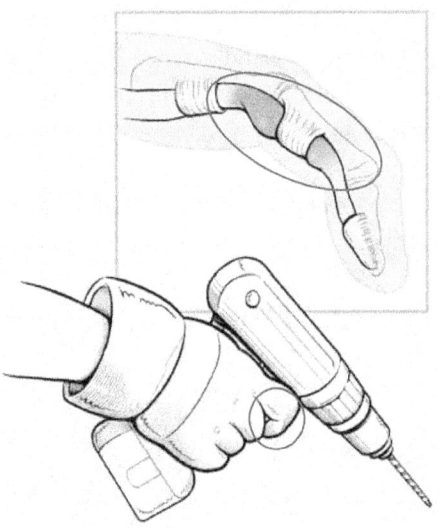

Trigger finger. Repeated pressure on a finger (such as when using the trigger on a power tool) may strain the tendon running to that finger, as well as the tendon covering This may cause discomfort or pain.

Epicondylitis. Forceful twisting motions may cause strain on your elbow tendons, causing discomfort or pain. This condition is called epicondylitis, also known as tennis elbow.

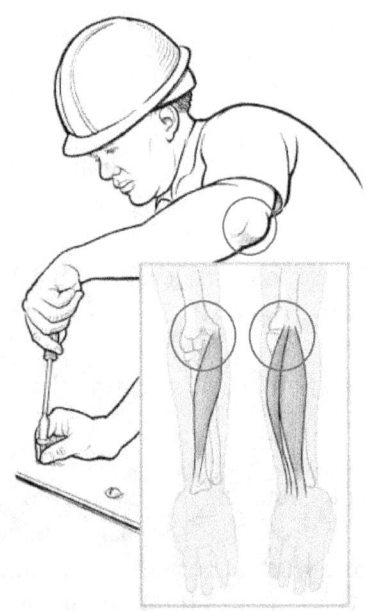

Hand-arm vibration syndrome (HAVS). Operating vibrating tools like needle guns, chipping hammers, and rotary hammer drills may lead to finger discoloration, tingling, and numbness. Gangrene is possible in the most severe cases.

Some Solutions

Hand-intensive work cannot be eliminated from construction, but it is possible to change how you do it so the work is easier on your body. Solutions are available that can reduce the level of stress on your hands, wrists, and arms. They may also reduce how often and how long your body is subjected to this stress. Many of the solutions can also eliminate other potential safety hazards and increase productivity.

The type of task and the site conditions will determine which solutions are best for you. A few possible solutions for specific hand-intensive tasks are explained in Tip Sheets #14–20.

General solutions for doing hand-intensive work with less risk of injury include:

Change materials or work processes. One of the most effective solutions may be to use materials, building components, or work methods that are less labor-intensive. For example, use lock nuts or button nuts on all-thread systems to reduce repeated hand-arm twisting and turning. An individual construction worker or subcontractor usually cannot make a decision like this. Certain changes may require the approval of the building owner, architect, engineer, or general contractor.

Change tools and/or equipment. If the work requires frequent intensive hand activity, you can often substitute a power tool for a manual tool. This will reduce the amount of hand force needed and the number of repeated movements, especially twisting motions. You will get the job done with less effort.

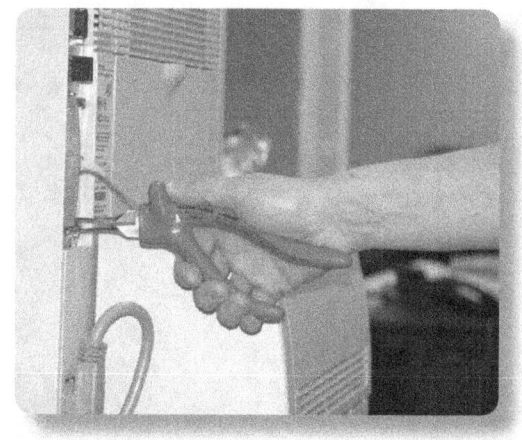

You can also use ergonomically improved tools. Select a tool that comes with a power grip, or add a power grip to existing tools. Compressible foam can be used to wrap a narrow tool handle and improve the grip. Some power tools have a large trigger that can be operated with multiple fingers so you don't constantly put all the stress on one finger. Also, look for power tools that meet stricter European hand-arm vibration requirements.

Offset handles can help keep the wrist straight

Choose the right tool for the job. For example, snips and other tools are available with features such as offset handles that can help keep your wrist straight on some types of jobs. Also, never use your hand to hammer or pound anything into place.

In a few cases, cost and site conditions may restrict the use of ergonomically improved tools.

Change work rules and provide training. Contractors can encourage the use of equipment like ergonomic tools. Site rules can limit the amount of time that workers do hand-intensive jobs without a break. A policy of providing ergonomics training may help workers more quickly identify potential problems and find effective solutions.

Ergonomic Hand Tools

The Problem

Using a conventional hand tool over and over can lead to muscle strain or even a serious injury like carpal tunnel syndrome or tendinitis. Using the wrong tool, or using a tool the wrong way, can strain your hand, wrist, forearm, shoulder, and neck.

One Solution

Use an "ergonomic" tool that fits the job. Many new tools are available that may help prevent muscle and joint injuries. However, some new tools advertised as "ergonomic" have not been carefully designed.

A tool can be considered "ergonomic" when it fits the task you do, fits your hand, allows a good grip, takes less effort, does not require you to work in an awkward position, does not dig into your fingers or hand, and is comfortable and effective. Remember that a tool designed for one task may put more stress on the hand or wrist when used for a different task. For example, needle-nose pliers work well for crimping electrical wire but should not be used for twisting.

How It Works

Here are some tips for selecting an ergonomic hand tool.

Handle. The handle should be non-slip, coated with soft material, and not have sharp edges. You may want to add a soft sleeve to the handle for a better and more comfortable grip. However, make sure the sleeve doesn't increase the handle diameter to more than

Avoid tools with finger grooves

Tool with soft grip and spring-loaded handle

Tool with offset handle can help keep wrist straight

two inches, which would make it harder to grip. Make sure the sleeve does not slip on the handle. If your task requires high force, pick a tool with a handle longer than the widest part of your hand. The end of the handle shouldn't press continually on your palm. Avoid tools with finger grooves on the grip. Grooves that do not fit your hand can put more pressure on your fingers, which can injure the finger tendons.

Wrist position. Pick a tool that keeps your wrist straight when you use it. A tool with a bent handle may work best if you are applying a horizontal force (in the same direction as your straight forearm and wrist). A tool with a straight handle may work best if you are applying an upward or downward force.

Handle diameter. For single-handle tools, if the task requires high force, handle diameter should be between 1-1/4 inches and 2 inches. For low-force tasks requiring precision or accuracy, handle diameter should be between 1/4 inch and 1/2 inch.

For double-handle tools, grip span for high-force tasks should be at least 2 inches when closed and no more than 3-1/2 inches when fully open. For low-force tasks requiring precision or accuracy, grip span should be at least 1 inch when closed and no more than 3 inches when fully open.

Pinching, gripping, or cutting tools. Choose a tool with a spring-loaded handle that automatically returns to an open position. If continuous high force is necessary, consider using a clamp, a grip, or locking pliers.

Benefits for the Worker and Employer

When you choose a tool that fits your task, you reduce your chance of developing an injury. You may also get the job done quicker and improve the quality of your work.

Approximate Cost

Many tool manufacturers now produce ergonomically improved hand tools. Often these are no more expensive than non-ergonomic tools.

For More Information

- Products related to this solution are described at *www.cpwr.com/simple.html*. Products also may be found on the internet using the following search terms: (type of tool) + "ergonomically designed."

- Local contractor tool and equipment suppliers or rental companies may be another source of information on products.

- For general information on this solution, check *www.cpwrconstructionsolutions.org* and *www.elcosh.org*. Other good information is available at:

    www.thomasnet.com (in the search box enter "tools: ergonomically designed")

    vendorweb.humantech.com/browse.asp

    www.advancedergonomics.com/product/tools.htm

TIP SHEET #15

Easy-Hold Glove for Mud Pans

The Problem

A mud pan full of drywall compound can weigh more than five pounds. Continuously gripping the pan can put a lot of stress on your hand, wrist, and forearm. If the mud pan is too wide for your hand, you have to squeeze its sides to hold it, putting more strain on your forearm muscles.

The smooth sides and bottom on a mud pan make it hard to grip with bare hands. Because of its weight, size, and smoothness, you must use a lot of hand force to hold the pan.

All these types of strain can tire your hand, wrist, arm, and forearm. If you do drywall work often and for long periods of time, the strain may lead to serious injuries.

One Solution

Use an **easy-hold glove** attached to the mud pan, which you can make yourself. The glove cuts down on the hand strength required to grip the pan. With the glove, all you need to do is balance the pan.

How It Works

The glove is bolted to the pan with a swivel mount. A bolt is welded to the bottom of the pan, and held by a nut inside the glove. This holds the pan in place. You never have to squeeze the pan. Because of the swivel mount, you can spin the pan in your hand

Problem: Holding mud pan without glove

Solution: Holding mud pan with glove

Glove assembly

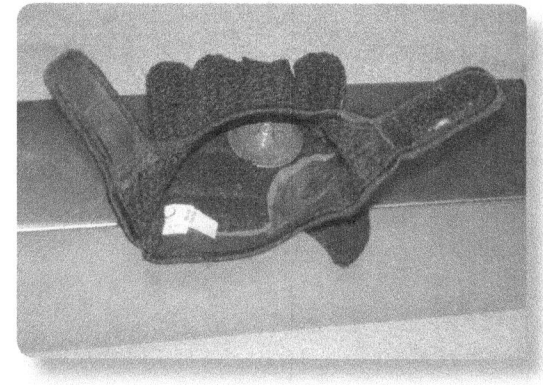

as needed. The glove fits snugly on your hand. A Velcro® strap holds it in place. The glove spreads the weight of the pan over your whole hand. You need only to open your hand wide to balance the pan. The pan swivels easily.

Benefits for the Worker and Employer

Finishers who used the easy-hold glove liked it and said it was great for long periods of coating drywall. They said that it cut down on their long-term fatigue and pain. Researchers found that wearing the glove led to a 25% reduction in grip force needed to hold the pan.

There are some drawbacks. Finishers said that it takes time to put the glove on and take it off. Other tasks, such as putting up tape, cannot be done while wearing the glove.

Approximate Cost

If you make your own easy-hold glove (see section below), the cost of materials is low. You can begin with a fingerless bicycle glove, available for $5–20.

To make your own: Use a tight-fitting glove with a stiff palm, such as a fingerless bicycle glove. Make sure the glove is not so tight that it affects blood circulation in your hand. But if the glove is too loose, you will end up gripping the mud pan tighter. A glove without fingers lets you use your own fingers more easily. However, it also makes the glove a little harder to take off, because the finger holes can get caught on your knuckles.

Fix the head of a small bolt to the bottom of the mud pan. You can weld the bolt or "glue" it using a metal-to-metal adhesive. Place a large (up to two-inch) washer on the bolt. Punch the bolt through the palm of the glove. Inside the glove, place another large washer over the bolt that sticks through. Tighten a lock nut over the bolt inside the glove. Cut off any part of the bolt that sticks above the nut and sand the bolt smooth. If the bolt still presses on the palm of your hand inside the glove, cover it with duct tape or another material.

For More Information

- Products related to this solution are described at *www.cpwr.com/simple.html*.

- Local contractor tool and equipment suppliers or rental companies may be another source of information on products.

- For general information on this solution, check *www.cpwrconstructionsolutions.org* and *www.elcosh.org*.

Power Caulking Guns

The Problem

Using a non-powered caulking gun requires high hand forces to pull the trigger. If you use these guns often and for long periods of time, you are at risk of stressing the soft tissues in your hand, wrist, and forearm. This can lead to serious muscle or joint injuries.

The more pressure your gun requires to pull the trigger (low mechanical advantage), the higher your chance of injury. You may have to use even more force when you apply thicker sealant material. If you have to bend your wrist or twist your forearm when pulling the trigger, you also increase your chance of injury.

Your chance of developing an injury increases if the gun you use has sharp edges or grooves on the trigger, or has a wide span between the trigger and the grip, forcing you to stretch your hand.

Problem: Manually squeezing caulking gun

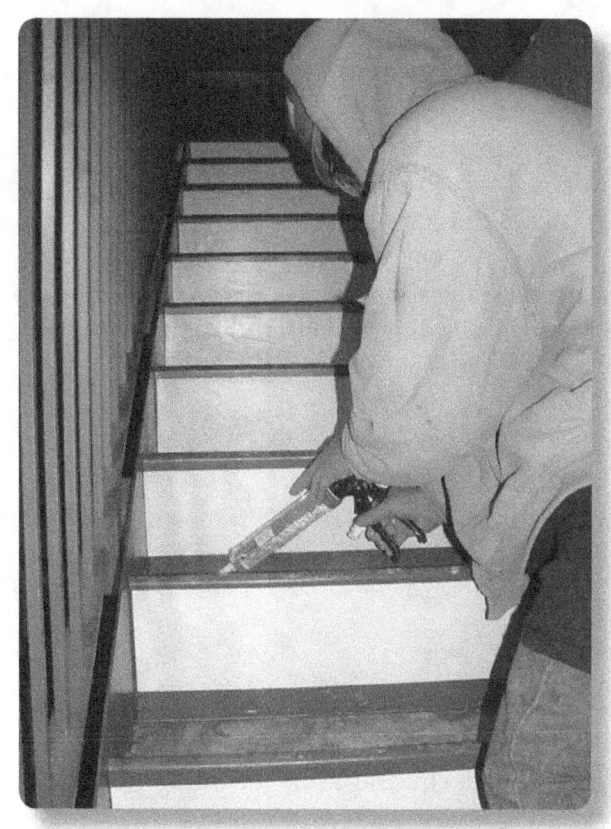

One Solution

Use a **power caulking gun**. These guns may be powered by a battery or may be pneumatic (powered by compressed air). With a power caulking gun you won't need to pull the trigger to apply the caulk or sealant. This will reduce the stress to your fingers, hand, and forearm.

Solution: Caulking adapter for drill

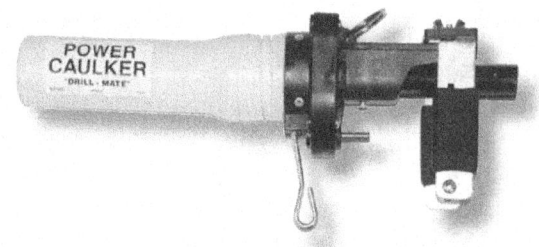

How It Works

A battery (cordless gun) or compressed air (pneumatic gun) provides the power to force the caulk from the gun. Some guns have a variable speed control to manage sealant flow. You do need access to a power source. This may be an electrical outlet or generator to charge the batteries in a cordless gun, or an air compressor to power a pneumatic gun. When you use a pneumatic gun, the hose may produce some resistance (or "drag") on the tool, possibly increasing the grip force necessary.

Cordless guns are available for tube, sausage, and bulk caulks. A 12-volt or higher cordless gun will produce 600+ lbs. of thrust force. Pneumatic guns will handle sausage and bulk caulks.

Another device available is a caulk-dispensing adapter that can be used with a cordless drill or variable speed AC drill. These adapters are available for 10 oz. and 30 oz. caulk cartridges.

Benefits for the Worker and Employer

Use of power caulking guns should result in less soft tissue fatigue, discomfort, and injury. Although power caulking guns are heavier than non-powered guns, they do not require high hand forces to get the caulk out of the gun.

One minor drawback is that pneumatic caulking guns are tethered to the air compressor with a hose, and the hose must be moved along with the gun.

Approximate Cost

Cordless caulking guns, with 12-volt battery and charger, range from $200–300, depending on size. Pneumatic caulking guns start around $20. Caulking attachments for cordless drills cost $45–50.

For More Information

- Products related to this solution are described at *www.cpwr.com/simple.html*. Products also may be found on the internet using the following search terms:

    Cordless Caulking Guns: "cordless caulk (or caulking) gun"

    Pneumatic Caulking Guns: "pneumatic caulk (or caulking) gun"

    Caulk-Dispensing Attachments for Drills: "power caulker drill attachment"

- Local contractor tool and equipment suppliers or rental companies may be another source of information on products.

- For general information on this solution, check *www.cpwrconstructionsolutions.org* and *www.elcosh.org*.

Reduced Vibration Power Tools

The Problem

From chainsaws to impact drills to soil compactors, some hand-held power tools can produce a lot of vibration. Certain tools generate vibration levels so high that they can damage the blood vessels and nerves in your hands.

The problem usually begins with numbness and tingling in your hands. After you have been exposed to vibration for a while, your fingertips may become discolored or white, especially when they get cold. You may also lose grip strength. In extreme cases, gangrene is possible. This injury is sometimes called "white finger" or "hand-arm vibration syndrome" (HAVS).

You are at greater risk of developing a vibration-related injury if you use vibrating power tools often or for long periods of time. These injuries can be prevented, but there is no cure once you have them.

One Solution

Use **reduced vibration power tools**. Many tools are now available that are designed to produce less vibration. They should always be used with full-fingered anti-vibration gloves that are certified to meet International Organization for Standardization (ISO) vibration standards (ISO 10819). Gloves that are not ISO-certified may not reduce vibration adequately, even when using a tool designed for low vibration.

Always keep your hands warm when using any vibrating tool, and use as light a hand grip as possible. Using a lighter grip helps reduce your exposure to vibration.

Problem: Some soil compactors can produce high vibration levels

Solution: Use reduced vibration tools with anti-vibration gloves such as these air bladder gloves

How It Works

Many manufacturers now offer reduced vibration tools. Although there are no regulations limiting vibration exposure in the U.S., there are limits in Europe and companies produce tools for both markets. The European limits have been adopted by the American National Standards Institute (ANSI) as recommended exposure limits (S2.70-2006).

The possible harm caused by operating a vibrating tool is related to the *level* of vibration and the *amount of time* the tool is used. The higher the vibration level, the shorter the time the tool can be used safely. In 2002, the European Union limited ISO frequency-weighted vibration exposure to an average of 5 meters per second per second (m/s^2) over any 8-hour period. For example, use of a tool with a high level of vibration ($10\ m/s^2$) would be limited to a shorter period (2 hours per day). For text of the European regulation search for "Directive 2002/44/EC" in any Internet search engine.

The National Institute for Working Life in Sweden lists vibration levels for hundreds of specific power tools on the web (go to *http://vibration.niwl.se/eng*). Vibration measurements are approximate, not absolute. They can vary depending on how the vibration is measured, how the tool is being used, and the condition of the tool. Also check users' manuals for vibration information.

If you know the vibration level of a tool, you can determine how much time you can safely use it. (This is called "trigger time.") There are several vibration exposure calculators on the web. If you enter a tool's vibration level (in m/s^2) the calculator will tell you the trigger time. Vibration exposure calculators can be used to help determine whether the tool you use produces too much vibration. Calculators can be found online by using the following search terms: "vibration exposure calculator."

Benefits for the Worker and Employer

Reduced vibration tools allows workers more "trigger time" with less risk of injury. Where trigger time limits are enforced by the employer, using reduced vibration tools may also increase productivity. Using anti-vibration gloves alone may not eliminate exposure to all harmful vibration.

Approximate Cost

Reduced vibration power tools are available for purchase and rental. Contact the tool manufacturer or representative for prices. Anti-vibration work gloves usually cost $40–50.

For More Information

- Products related to this solution are described at *www.cpwr.com/simple.html*. Products also may be found on the internet using the following search terms: "low vibration tools."

- Local contractor tool and equipment suppliers or rental companies may be another source of information on products.

- For general information on this solution, check *www.cpwrconstructionsolutions.org* and *www.elcosh.org*. Information, including additional vibration exposure calculators, is also available from the Canadian Center for Occupational Health and Safety (CCOHS) at *www.ccohs.ca*.

Power Cleaning and Reaming with a Brush

The Problem

Frequent use of a wire brush to clean or ream pipes, grates, and other building materials can strain your hands, wrists, forearms, and elbows. Using the brush may be light work, but you must bend your wrist and use fast pulling, pushing, or rotating motions. If you do this work often, you can be at risk of a serious muscle or joint injury.

Your chance of developing a serious injury increases when you have to apply high hand force to the brush or use a pinch grip to hold it. If you wear loose-fitting, thick gloves, the brush may be harder to hold and require more force.

Problem: Brushing copper tube by hand

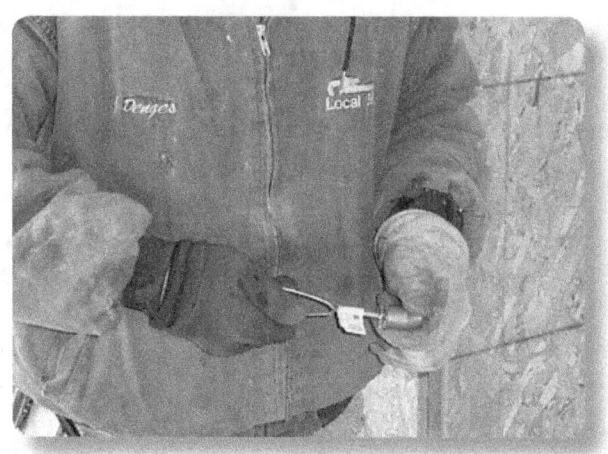

One Solution

Place the wire brush in the chuck of a battery-powered or corded **screw gun or screwdriver.** The power tool will eliminate the repeated hand, wrist, and forearm motions and may improve your grip.

Solution: Using wire brush with power driver

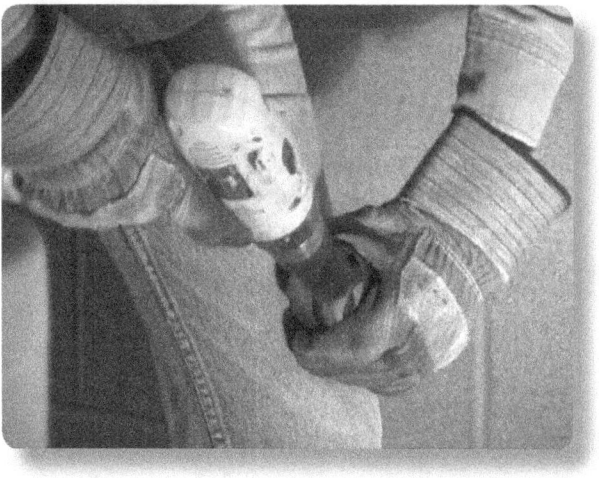

How It Works

The power tool rotates the brush for you. Choose a power tool with a soft, non-slip handle (plastic or rubber covering). The handle should be large enough to fit your whole hand. It should not have sharp edges or ridges. You will need less hand strength to grip this type of handle than to grip a brush. The smaller handles on manual wire brushes must be gripped with a few fingers, rather than with the full hand. This requires more muscle force.

There will be less localized pressure on your hand because the handle won't dig into your skin. There also will be fewer rapid movements of your hand, wrist, and forearm. The rotation of the power tool replaces the motions necessary with the manual wire brush.

Many power tools can help keep your wrist straight. Some have handles that change from in-line to pistol-grip (two-way handles). These prevent awkward wrist positions because the tool bends—not your wrist. Or, depending on the location of the work, you can use a power tool with either a pistol grip or in-line handle, whichever best reduces wrist bending in each situation.

Benefits for the Worker and Employer

Using a power tool with the wire brush should result in less strain on your hand, wrist, forearm, and elbow. It will also improve productivity because the work is faster than using a brush by hand.

You do need access to a power source. This may be an electrical outlet or generator to run a corded tool or to charge the batteries in a cordless one.

Approximate Cost

Professional-quality tools can be purchased at most hardware, home improvement, and commercial building supply stores. Prices vary and you should shop around before purchasing one. If you plan to use the tool often, consider purchasing a heavy-duty contractor or professional-quality model. For professional models, a corded screw gun runs $125–150. A battery-powered screw gun (14.4–18 volts) is $180–250. A battery-powered screwdriver (2.4–3.4 volts) is $100–125. Wire brushes are available as accessories from many power tool manufacturers.

For More Information

- Products related to this solution are described at *www.cpwr.com/simple.html*.

- Local contractor tool and equipment suppliers or rental companies may be another source of information on products.

- For general information on this solution, check *www.cpwrconstructionsolutions.org* and *www.elcosh.org*.

Snips for Cutting Sheet Metal

The Problem

Cutting sheet metal with snips takes a lot of hand force. You often need to work with your wrist in an awkward position. If you do this work often or for long periods of time, you may experience hand or wrist pain. Eventually you may develop a serious injury.

Using the wrong snip for the job increases your chance of injury. Snips come in many shapes and sizes. Manufacturers produce different snips for specific tasks and specific workers. If you use left-cut snips to do a right-cut task, your hand and wrist will be in a stressful position and you will have to use more force. If you cut sheet metal that is thicker than the snip manufacturer recommends, more force will also be necessary. If you use dull snips, they will make your work even harder.

One Solution

Use the **right size and type of snip** for the task. New types of snips are available that may fit your hand better, keep your wrist straighter, and require less hand force.

Any snip you use should be sharp and tight. Do not use dull or damaged snips. Where necessary use compound snips, which provide much more power. Some compound snips will increase your hand strength by 12 times. Electric snips are usually best when you need to make many cuts.

Most snips are made for cutting soft metal only. For hard metal, you should use other cutting tools designed for that purpose.

Problem: All-purpose snips are not suitable for every job

One solution: Tapping into duct using upright snip

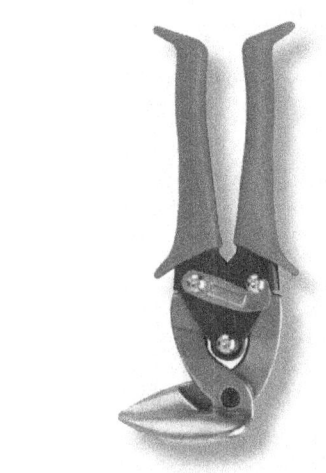

How It Works

Improvements found in the newer snips include less space between the handles, soft grips and curved handles. Using a curved handle can help keep your wrist straight. A soft grip lowers the pressure on your hand and fingers. When the space between handles is less, you may be able to get a better grip on the tool. An upright snip can help keep your wrist straighter when working in confined areas or overhead.

No pair of snips will work well for every task. Decide what the job requires and select the correct snips. Manufacturers make specific snips for left and right cuts, straight cuts, and wide or tight curved cuts. There are specific snips for different gauges of sheet metal. Different snips are made to fit left- and right-handed workers.

Pay attention to the manufacturer's specifications. Manufacturers color-code snip handles for the type of cut they make— yellow for straight, green for right, and red for left. Use snips that fit both your hand and the job you do, either left-handed or right-handed. Always wear eye protection when using snips.

Benefits for the Worker and Employer

Choosing the correct snips should make the job easier. Your hand and wrist should be less tired, and less likely to be injured. You also should be able to get the job done faster.

Approximate Cost

The best snip for the job should not cost much more than other snips. Newer snips are often between $10–40.

For More Information

- Products related to this solution are described at *www.cpwr.com/simple.html*. Products also may be found on the internet using the following search terms: "aviation snips" + "ergonomic design."

- Local contractor tool and equipment suppliers or rental companies may be another source of information on products.

- For general information on this solution, check *www.cpwrconstructionsolutions.org* and *www.elcosh.org*.

TIP SHEET #20

Quick-Threading Lock Nuts

The Problem

When you tighten a standard lock nut around the thread on a long rod, you have to twist your hand, wrist, and forearm over and over. Making these twisting movements can strain the muscles and tendons in your hand, wrist, and elbow. The strain can become more serious if you do this work a lot and you repeat the same movements for a long period of time. You can eventually develop pain and even a serious injury.

Your chance of injury depends on the amount of finger pressure you use to hold the nut, the distance the nut is threaded, and the number of nuts threaded. Working in positions where you have to reach above your shoulders to thread the nut increases your chance of injury.

Problem: Tightening conventional nut on all-thread

One Solution

Use a **quick-threading lock nut**. Depending on the type of nut, these can either snap onto an all-thread rod at any position, or slide up and down the rod freely. They eliminate the repeated hand, wrist, forearm, and elbow twisting. They can also reduce the time you spend working above your shoulders because they go on faster.

Solution: Two piece slip-on lock nut (top) and button lock nut (bottom)

How It Works

There are two types of quick-threading lock nuts—two piece lock nuts and button lock nuts.

When using the two piece lock nut, you twist the two sections of the nut apart until the slot is open, then place the nut on the all-thread rod where you need it. Then twist the two sections together again until the slot is closed and the sections are snug against each other. Finally, tighten the nut with a wrench until the openings on the two sections face in opposite directions.

When using the button lock nut, you first push the button to loosen the nut from the thread. Then slide the nut to the spot you want. Let go of the button to make the nut re-connect with the thread. Then tighten it as you would any nut.

Benefits for the Worker and Employer

Using quick-threading lock nuts should reduce the strain on your hand, wrist, and forearm. It should therefore lessen the chance of pain and musculoskeletal injury. Because you spend less time working above your shoulders, there is also less overall strain on your shoulders, neck, and back during a shift.

In addition, these nuts should lead to a gain in productivity because they take less time. They make certain kinds of work much easier, such as threading nuts in locations that are difficult to access. Quick-threading lock nuts may not be appropriate for all jobs, and their use may require the approval of the building owner, architect, engineer, or general contractor.

Approximate Cost

The two piece lock nuts start at $2–3 each, depending on diameter. Metric sizes are available. The button lock nuts start at approximately $6 each.

For More Information

- Products related to this solution are described at *www.cpwr.com/simple.html*.

- Local contractor tool and equipment suppliers or rental companies may be another source of information on products.

- For general information on this solution, check *www.cpwrconstructionsolutions.org* and *www.elcosh.org*.

Glossary

ANSI

American National Standards Institute. ANSI is a private, non-profit membership organization that coordinates voluntary standards in many fields. ANSI encourages the private sector and government to reach agreement on the need for standards and establish priorities.

Arthritis

Inflammation of a joint or joints in the body.

Awkward position

See *Awkward posture*.

Awkward posture

Deviation from the natural or "neutral" position of a body part. A neutral position is one that puts minimal stress on the body part. Awkward postures typically include reaching above or behind, twisting, bending forward or backward, pinching, squatting, and kneeling. Working frequently in awkward postures can cause fatigue, pain, and musculoskeletal injury.

Body mechanics education

Education that emphasizes how best to align the musculoskeletal system during work and other activities to reduce abnormal joint stress, muscle strain, and fatigue.

Bursa

Small, flat, fluid-filled sacs located in those areas of the body where repeated pressure is exerted during movement of body parts, such as the shoulder, elbow, and knee. Bursa allow these body parts to move more easily.

Bursitis

Inflammation or irritation of the bursa, resulting in swelling, stiffness, and pain.

Carpal tunnel

An opening inside the wrist through which the median nerve and several tendons pass. The tunnel is formed by the wrist bones and a dense ligament.

Carpal tunnel syndrome (CTS)

A condition in which there is pressure on the median nerve in the carpal tunnel. The nerve gets squeezed when the tendons swell. Symptoms can include pain, tingling, or numbness in the hand, wrist, or arm. These symptoms are often felt at night.

Cartilage

Thick, white connective tissue attached to the surfaces of bones where they contact other bones, forming a low-friction cushion. It is structurally more rigid than a tendon.

Cervical vertebrae

Seven small irregular bones in the neck that support and allow head movement.

Contact stress

Pressure on one specific area of the body (such as the forearm or sides of the fingers) that can inhibit nerve function and blood flow in that area. It is caused by continuous or repeated contact with hard or sharp objects such as table edges or unpadded, narrow tool handles.

Cumulative trauma disorder (CTD)

An injury that develops over a period of time because of repeated stress on a specific body part, such as the back, hand, wrist, or forearm. Muscles and joints are stressed, tendons are inflamed, nerves are pinched, and/or the flow of blood is restricted. Similar to *Repetitive stress injury.*

Discs

See *Intervertebral discs.*

Disorder

A medical condition in which some body function does not work as it should.

Epicondylitis

An inflammation of the tendons at the elbow. It is also called "tennis elbow" (lateral or outside part of the elbow), or "golfer's elbow" (medial or inside part of the elbow).

Ergonomics

The science of fitting workplace conditions and job demands to the capabilities of workers' bodies.

Ergonomics program

A systematic process, often spelled out in writing, for identifying, analyzing, and controlling ergonomic hazards at a particular workplace.

Fatigue

A condition that results when the body cannot provide enough energy for the muscles to perform a task.

Force

The amount of physical effort needed to do a task.

Gangrene

Death of body tissue as a result of a loss of blood flow to the area.

Grip force

Physical force applied by the hand when holding or gripping an object.

Hand-arm vibration

Vibration (generally from a hand tool) that goes through the hand and can travel to the arm and other areas of the body.

Hand-arm vibration syndrome (HAVS)

Numbness, tingling, and whitening of the fingers due to exposure to hand-arm vibration. It is often caused by using vibrating hand tools frequently or for long periods of time. It involves blood vessel damage, such as closure of the digital (finger) arteries.

Herniated disc

A condition where the soft inner part of an intervertebral disc pushes out through a tear in the disc.

In-line grip

A hand tool handle which is straight.

Inflammation

A protective response of the body to infection and injury. Symptoms may include tissue swelling, redness, pain, and a feeling of warmth.

Intervertebral discs

Discs that sit between the bones of the spinal column (vertebrae) in the back and neck. The discs act as cushions or "shock absorbers" between the bones. Discs have a strong outer wall and a soft inner gel.

ISO

The International Organization for Standardization. This is a non-governmental organization, a network consisting of the national standards institutes of 157 countries.

Joint

The area where two bones are attached to allow body movement. A joint is usually formed of ligaments and cartilage.

Ligaments

Strong rope-like fibers that connect one bone to another to form a joint.

Manual material handling

Lifting, carrying, and moving materials without the help of mechanical equipment.

Median nerve

The main nerve passing through the carpal tunnel in the wrist.

Muscle force

Physical force applied with the muscles.

Musculoskeletal disorders (MSDs)

A group of conditions that involve the nerves, tendons, muscles, and supporting structures such as intervertebral discs. The various conditions can differ in severity from mild symptoms once in a while to severe chronic and disabling disorders. Examples include carpal tunnel syndrome, tenosynovitis, tension neck syndrome, and low back pain.

Musculoskeletal system

The soft tissues and bones in the body. The parts of the musculoskeletal system are bones, muscles, tendons, ligaments, cartilage, nerves, and blood vessels.

Nerves

Cordlike fibers that carry the signals controlling body movement and allowing senses like sight and touch to work.

Neutral body posture

The natural position of body parts, the best position to minimize stress. For example, when standing, the head should be aligned over the shoulders, shoulders aligned over hips, hips aligned over ankles, and elbows at the side of the body.

Neutral position
> See *Neutral body posture*.

NIOSH
> National Institute for Occupational Safety and Health. NIOSH, part of the Centers for Disease Control and Prevention (CDC) under the Department of Health and Human Services, is the federal government agency with a mandate to conduct and fund occupational safety and health research and training.

OSHA
> Occupational Safety and Health Administration. OSHA is a federal government agency, part of the U.S. Dept. of Labor, whose mission is to help prevent workplace injuries and protect the health of workers. OSHA adopts and enforces workplace health and safety standards.

Pistol grip
> A hand tool handle which resembles the handle of a pistol and is typically used when the tool axis must be horizontal.

Power grip
> A grasp in which the hand wraps completely around a handle. The handle runs parallel to the knuckles and protrudes on either side.

Repetitive stress injury (RSI)
> An injury caused by working in the same awkward position, or repeating the same stressful motions, over and over. This is one type of *Musculoskeletal disorder*.

Risk factor
> An action and/or condition that may cause an injury or illness, or make it worse. Examples related to ergonomics include forceful exertion, awkward posture, and repetitive motion.

Rotator cuff
> The main source of stability and mobility for the shoulder. Four muscles and their tendons make up the rotator cuff. They wrap around the front, back, and top of the shoulder joint. They rotate the arm inward, outward, and away from the side.

Rotator cuff tear
> A tear in the rotator cuff caused by stress on the shoulder. A tear can make routine activities difficult and painful.

Rotator cuff tendinitis
> The most common shoulder disorder, involving inflammation, pain, and often swelling in one or more tendons of the rotator cuff. It is sometimes called "pitcher's shoulder."

Ruptured disc
> See *Herniated disc*.

Soft tissues
> Tissues that connect, support, or surround other structures and organs of the body.

Sprain
> Overstretching or overexertion of a ligament, resulting in a tear or rupture of the fibers in the ligament.

Strain

An injury caused by a muscle, tendon, or ligament stretching.

Stress

Demand (or "burden") on the human body caused by something outside of the body, such as a work task, the physical environment, work-rest schedules, and social relationships.

Tendinitis

Inflammation, fraying, or tearing of tendon fibers, resulting in pain and sometimes swelling.

Tendon

Tough rope-like material that connects the muscles to the bones. Tendons transfer forces and movements from the muscles to the bones. Tendons do not stretch, and excessive force or twisting may cause them to tear or fray like a rope.

Tenosynovitis

Inflammation of the lining of the sheath that surrounds a tendon. The wrists, hands, and feet are the areas commonly affected, although tenosynovitis may occur in any tendon sheath.

Tension neck syndrome (TNS)

Fatigue, stiffness, tenderness, swelling, weakness, or pain in the neck or shoulder area, or headache radiating from the neck. It is caused by strain on various neck and shoulder muscles, often from long periods of looking upward. The trapezius muscle is particularly affected and may develop a "knot."

Thoracic outlet syndrome

A cumulative trauma disorder of the nerves and blood vessels of the shoulder and upper arm. Symptoms are numbness in the fingers or arm. The pulse in the affected area may weaken.

Trapezius muscle

A large, thin muscle that runs from the upper back through the shoulder area to the neck. Straining this muscle can cause tension neck syndrome.

Trigger finger

A common term for tendinitis or tenosynovitis that causes painful locking of the finger(s) while flexing them. It can be caused by repeated pressure on a finger, such as when using the trigger on a power tool.

Trigger time

The length of time a person can safely use a vibrating power tool, based on its vibration level.

Whole body vibration (WBV)

Working conditions that involve sitting, standing, or lying on a vibrating surface. Excessive exposure may contribute to back pain.

Work-related musculoskeletal disorder (WMSD)

A musculoskeletal disorder caused or made worse by the work environment. WMSDs can cause severe symptoms such as pain, numbness, and tingling; reduced productivity; lost time from work; temporary or permanent disability; loss of motion; inability to perform job tasks; and an increase in workers' compensation costs.

Definitions adapted in part from ergonomics materials provided by NIOSH, Cornell University, the Virginia Polytechnic Institute and State University, and the Washington State Dept. of Labor and Industries.

SIMPLE SOLUTIONS
References

Why This Booklet?

Center to Protect Workers' Rights [2002]. Construction Chart Book, 2nd edition. Silver Spring, MD: CPWR. [www.cdc.gov/elcosh/docs/d0100/d000038/sect41.html]. Date accessed: July 2006.

Cook TM, Rosecrance JC, Zimmerman CL [1996]. The University of Iowa construction survey. Washington, DC: Center to Protect Workers' Rights, Report No. E1-96.

Schneider S [1995]. Ergonomics. Implement Ergonomic Interventions in Construction. Applied Occupational and Environmental Hygiene *10*:822–824.

Oh, My Aching Body!

Cook TM, Rosecrance JC, Zimmerman CL [1996]. The University of Iowa construction survey. Washington, DC: Center to Protect Workers' Rights, Report No. E1-96.

NIOSH [2006]. Proceedings of a meeting to explore the use of ergonomic interventions for the mechanical and electrical trades. Cincinnati, OH: U.S. Department of Health and Human Services, Centers for Disease Control and Prevention, National Institute for Occupational Safety and Health, DHHS (NIOSH) Publication No. 2006-119.

Silverstein B, Kalat J [1998]. Work-related disorders of the back and upper extremity in Washington State, 1989-1996. Olympia, WA: SHARP Program, Washington State Department of Labor and Industries, TR 40-1-1997.

Simple Solutions for Floor and Ground-Level Work — Introduction

Haslegrave CM, Tracy MF, Corlett EN [1997]. Strength capability while kneeling. Ergonomics *40*(12):1363-1379.

Kirkesov-Jensen L, Eenberg E [1996]. Occupation as a risk factor for knee disorders. Scandinavian Journal of Work Environment and Health *22*:165-175.

Maher CG [2000]. A systematic review of workplace interventions to prevent low back pain. Australian Journal of Physiotherapy *46*:259-269.

Manninen P, Heliövaara M, Riihimäki S [2002]. Physical workload and the risk of severe knee osteoarthritis. Scandinavian Journal of Work Environment and Health *29*(1):25-32.

National Research Council and Institute of Medicine (NRC/IOM) [2001]. Musculoskeletal disorders and the workplace. Washington, DC: National Academy Press.

Pope MH, Koh KL, Magnusson ML [2002]. Spine ergonomics. Annual Review of Biomedical Engineering *48*:49-68.

Ritz B, Brunnholzl K [1988]. Knee-joint lesions of pipe-fitters and welders employed by the public water and gas works. In: Hogstedt C, Rueterwall C (eds.), Progress in occupational epidemiology. Proceedings of the Sixth International Symposium on Epidemiology in Occupational Health in Stockholm, Sweden, 16-19 August 1988. Amsterdam: Elsevier Science Publishers B.V., pp.227-230.

Seidler A, Bolm-Audorff U, Heiskel H, Henkel N, Roth-Küver B, Kaiser U, Bickeböller R, Willingstorfer WJ, Beck W, Elsner G [2001]. The role of cumulative physical work load in lumbar spine disease: risk factors for lumbar osteochondrosis and spondylosis associated with chronic complaints. Occupational and Environmental Medicine *58*:735-746.

Solomonow M, Baratta RV, Banks A, Freudenberger C, Zhou BH [2003]. Flexion-relaxation response to static lumbar flexion in males and females. Clinical Biomechanics *18*:273-279.

Tip Sheet #1. Fastening Tools that Reduce Stooping

Bernold LE, Lorenc SJ, Davis ML [2001]. Technological intervention to eliminate back injury risks for nailing. Journal of Construction Engineering and Management *127*(3):245-250.

Hess JA, Kincl L, Albers J [2006]. Evaluation of a tool extension to reduce low back injury in carpenters. Proceedings of the International Ergonomics Association 2006 Congress, The Netherlands, July 10-14, 2006.

Tip Sheet #2. Motorized Concrete Screeds

Albers J, Russell S, Stewart K [2004]. Concrete leveling techniques: A comparative ergonomics assessment. Proceedings of the Human Factors and Ergonomics Society 48th Annual Meeting, New Orleans, LA, September 20-24, 2004.

Goldsheyder D, Weiner SS, Nordin M, Hiebert R [2004]. Musculoskeletal symptom survey among cement and concrete workers. Work *23*(2):111-121.

Tip Sheet #3. Rebar-Tying Tools

Albers JT, Hudock SD [2007]. Biomechanical assessment of three rebar tying techniques. International Journal of Occupational Safety and Ergonomics *13*(3):279-289.

Albers J, Hudock S, Kong YK [2005]. NIOSH Health Hazard Evaluation Report, Genesis Steel Services, Inc. Cincinnati, OH: U.S. Department of Health and Human Services, Centers for Disease Control and Prevention, National Institute for Occupational Safety and Health, HETA 2003-0146-2976. [http://www.cdc.gov/niosh/hhe/reports/pdfs/2003-0146-2976.pdf]

Dababneh AJ, Waters TR [2000]. Ergonomics of rebar tying. Applied Occupational and Environmental Hygiene *15*(10):721-727.

Forde M [2002]. Reinforcing ironwork: PATH (posture, activity, tools, handling) analysis. Lowell, MA: Construction Occupational Health Program, Department of Work Environment, University of Massachusetts Lowell. Technical Report T-61. [www.uml.edu/Dept/WE/COHP]. Date accessed: December 2004.

Vi P [2003]. Reducing risk of musculoskeletal disorders through the use of rebar-tying machines. Applied Occupational and Environmental Hygiene *18*(9):649-654.

Vi P [2005]. Promoting early return to pre-injury job using a rebar-tying machine. Journal of Occupational and Environmental Hygiene *2*:D34-D37.

Tip Sheet #4. Kneeling Creepers

Jensen LK, Mikkelsen S, Loft IP, Eenberg W [2000]. Work-related knee disorders in floor layers and carpenters. Journal of Occupational and Environmental Medicine *42*(8):835-842.

Kivimäki J, Riihimäki H, Hänninen K [1992]. Knee disorders in carpet and floor layers and painters. Scandinavian Journal of Work Environment and Health *18*:310–316.

Tip Sheet #5. Adjustable Scaffolding for Masonry Work

Breithaupt J [2005]. A scaffold by any other name. Masonry *44*(4).

Breithaupt J [2005]. Saving the day … Each and every day. Masonry *43*(3).

de Jong AM, Vink P, De Kroon JC [2003]. Reasons for adopting technological innovations reducing physical workload in bricklaying. Ergonomics *46*(11):1091-1108.

Entzel P, Albers JT, Welch L [2007]. Ergonomic best practices for masonry construction. Applied Ergonomics *38* (2007): 557–566.

Fletcher LT [1973]. Masonry productivity (Thesis). Austin, TX: University of Texas at Austin, College of Engineering, Center for Building Research.

Gilbreth, FB [1909]. Bricklaying systems. New York: Myron Clark.

Jorgensen K, Jensen BR, Kato M [1991]. Fatigue development in the lumbar paravertebral muscles of bricklayers during the working day. International Journal of Industrial Ergonomics *8*:237-245.

Luttmann A, Jager M, Laurig W [1996]. Task analysis and electromyography for bricklaying at different wall heights. International Journal of Industrial Ergonomics *8*:247-260.

Sak E [2003]. Adjustable scaffolding safety benefits. Masonry Construction Magazine *42*(7).

Suprenant BA [1990]. Tower scaffolding increases productivity 20%. Masonry Construction Magazine *34*(7):20-23.

University of Texas, Austin [1974]. Findings of masonry productivity research. Austin, TX: Contract H-1470, U.S. Department of Housing and Urban Development.

Urlings IJM, Wortel E [1991]. Implementation of an ergonomically improved adjustable height platform in the Dutch building and construction industry. Proceedings of the 11th Triennial Congress of the International Ergonomics Association, France, 2006.

van der Molen HF, Grouwstra1 R, Kuijer P, Sluiter JK, Frings-Dresen MHW [2004]. Efficacy of adjusting working height and mechanizing of transport on physical work demands and local discomfort in construction work. Ergonomics *47*(7):772-783.

Vink P, Koningsveld EAP [1990]. Bricklaying: a step by step approach to better work. Ergonomics *33*(3):349-352.

Simple Solutions for Overhead Work — Introduction

National Research Council and Institute of Medicine (NRC/IOM) [2001]. Musculoskeletal disorders and the workplace. Washington, DC: National Academy Press.

Welch LS, Hunting KL, Kellogg J [1995]. Work-related musculoskeletal symptoms among sheet metal workers. American Journal of Industrial Medicine *27*(6):783-791.

Tip Sheet #6. Bit Extension Shafts for Drills and Screw Guns

Anton D, Shibley LD, Fethkes NB, Hess J, Cook TM, Rosecrance J [2001]. The effect of overhead drilling position on shoulder moment and electromyography. Ergonomics *44*(5):489-501.

Tip Sheet #7. Extension Poles for Powder-Actuated Tools

Wos H, Lindberg J, Jakus R, Norlander S [1992]. Evaluation of impact loading in overhead work using a bolt pistol support. Ergonomics *35*(9):1069-1079.

Tip Sheet #8. Spring-Assisted Drywall Finishing Tools

Pan CS, Chiou SS, Hsiao H, Becker P, Akladios M [2000]. Assessment of perceived traumatic injury hazards during drywall taping and sanding. International Journal of Industrial Ergonomics *25*:621-631.

Washington State Department of Labor and Industries [2002]. Wallboard: ergonomics demonstration project. [www.lni. wa.gov/wisha/ergo/demofnl/wallboard_fnl.pdf]. Date accessed: September 2005.

Tip Sheet #9. Pneumatic Drywall Finishing Systems

Construction Safety Association of Ontario [2004]. Ergonomic and hygiene interventions to improve the health and safety of drywall finishing workers. [www.wsib.on.ca/wsib/wsibsite.nsf/public/researchergonomichygienedrywallworkers]. Date accessed: September 2005.

Pan CS, Chiou SS, Hsiao H, Becker P, Akladios M [2000]. Assessment of perceived traumatic injury hazards during drywall taping and sanding. International Journal of Industrial Ergonomics *25*:621-631.

Washington State Department of Labor and Industries [2002]. Wallboard: ergonomics demonstration project. [www.lni. wa.gov/wisha/ergo/demofnl/wallboard_fnl.pdf]. Date accessed: September 2005.

Simple Solutions for Lifting, Holding, and Handling Materials — Introduction

Dempsey PG, Hashemi L [1999]. Analysis of workers' compensation claims associated with manual materials handling. Ergonomics *42*(1):183-195.

Gallagher S, Hamrick CA, Cornelius KM, Redfern MS [2001]. The effects of restricted workspace on lumbar spine loading. Occupational Ergonomics *2*(4):201-213.

Hess J, Hecker S [2003]. Stretching at work for injury prevention: Issues, evidence, and recommendations. Applied Occupational and Environmental Hygiene *18*(5):331-338.

Holmström EB, Lindell J, Moritz U [1992]. Low back and neck/shoulder pain in construction workers: Occupational workload and psychosocial risk factors. Part 2: Relationship to neck and shoulder pain. Spine *17*(6):672-677.

Latza U, Pfahlberg A, Gefeller O [2002]. Impact of repetitive manual materials handling and psychosocial work factors on the future prevalence of chronic low-back pain among construction workers. Scandinavian Journal of Work Environment and Health *28*(5):314-323.

National Research Council and Institute of Medicine (NRC/IOM) [2001]. Musculoskeletal disorders and the workplace. Washington, DC: National Academy Press.

NIOSH [1994]. Applications manual for the revised NIOSH lifting equation. Cincinnati, OH: U.S. Department of Health and Human Services, Centers for Disease Control and Prevention, National Institute for Occupational Safety and Health, DHHS (NIOSH) Publication No. 94-110.

Pope MH, Koh KL, Magnusson ML [2002]. Spine ergonomics. Annual Review of Biomedical Engineering *48*:49-68.

Waters TR, Putz-Anderson V, Garg A [1993]. Revised NIOSH equation for the design and evaluation of manual lifting tasks. Ergonomics *36*(7):749-776.

Tip Sheet #10. Lightweight Concrete Block

Anton D, Rosecrance JC, Gerr F, Merlino LA, Cook TM [2005]. Effect of concrete block weight and wall height on electromyographic activity and heart rate of masons. Ergonomics *48*(10):1314-1330.

Brouwer J, Bulthuis BM, Begemann-Meijer M [1991]. The workload of gypsum bricklayers: the effect of lowering the mass and reducing the size of a gypsum brick. In: Queinnec Y, Daniellou F (eds.), Designing for everyone: Proceedings of the Eleventh Congress of the International Ergonomics Association. London: Taylor and Francis.

de Looze MP, Visser B, Houting I, van Rooy MA, van Dieen JH, Toussaint HM [1996]. Weight and frequency effect on spinal loading in a bricklaying task. Journal of Biomechanics *29*(11):1425-1433.

Entzel P, Albers JT, Welch L [2007]. Ergonomic best practices for masonry construction. Applied Ergonomics *38* (2007): 557–566.

Expanded Shale and Clay Institute. High performance concrete masonry: Information Sheet 3650.4 for mason contractors. [www.smartwall-systems.org]. Date accessed: February 2006.

Lochonic L [2003]. Lightweight CMU: A weight off our shoulders. Livonia, MI: Masonry Institute of Michigan, The Story Pole, *34*(4). [www.escsi.org]. Date accessed: October 2005.

Zellers K, Simonton K [1997]. An optimized lighter-weight concrete masonry unit: Biomechanical and physiological effects on masons. Olympia, WA: SHARP Program, Washington State Department of Labor and Industries.

Tip Sheet #11. Pre-Blended Mortar and Grout Bulk Delivery Systems

Entzel P, Albers JT, Welch L [2007]. Ergonomic best practices for masonry construction. Applied Ergonomics *38* (2007): 557–566.

Goldsheyder D, Nordin M, Weiner SS, Hiebert R [2002]. Musculoskeletal symptom survey among mason tenders. American Journal of Industrial Medicine *42*(5):384-396.

Schierhorn C [1996]. Dispensing preblended mortar into conventional mixers. The Aberdeen Group, Masonry Construction, Publication #M960369. [ftp://imgs.ebuild.com/woc/M960369.pdf]. Date accessed: August 2005.

Tip Sheet #12. Skid Plates to Move Concrete-Filled Hoses

Ahn K, Paquet VL, Buchholz B [2000]. Ergonomic assessment of the concrete pouring operation during highway construction. 128th Annual Meeting of American Public Health Association, Boston, MA. [apha.confex.com/apha/128am/techprogram/paper_13287.htm]. Date accessed: September 2005.

Hess JA, Hecker S, Weinstein M, Lunger M [2004]. A participatory ergonomics intervention to reduce risk factors for low-back disorders in concrete laborers. Applied Ergonomics *35*(5):427-441.

Occupational and Industrial Orthopaedic Center [2003]. Ergonomics working for cement and concrete construction laborers. [www.lhsfna.org]. Date accessed: September 2005.

Tip Sheet #13. Vacuum Lifters for Windows and Sheet Materials

Schwind GF [1994]. The ergonomics of below-the-hook lifters. Material Handling Engineering *49*(4):77-81.

Simple Solutions for Hand-Intensive Work — Introduction

Chao A, Kumar AI,. Emery C, Nagaajaao K, You H [2000]. An ergonomic evaluation of Cleco pliers. Proceedings of the IEA 2000/HFES 2000 Congress, USA, July 29 - August 4, 2000.

Keyserling WM [2000]. Workplace risk factors and occupational musculoskeletal disorders. Part 2: A review of biomechanical and psychophysical research on risk factors associated with upper extremity disorders. American Industrial Hygiene Association Journal *61*(2):231-243.

Marras WS, Schoenmarklin RW [1993]. Wrist motions in industry. Ergonomics *36*(4):342-351.

National Research Council and Institute of Medicine (NRC/IOM) [2001]. Musculoskeletal disorders and the workplace. Washington, DC: National Academy Press.

NIOSH [1997]. Musculoskeletal disorders and workplace factors, 2nd edition. Cincinnati, OH: U.S. Department of Health and Human Services, Centers for Disease Control and Prevention, National Institute for Occupational Safety and Health, DHHS (NIOSH) Publication No. 97-141.

Rosecrance JC, Cook TM, Anton DC, Merlino LA [2002]. Carpal tunnel syndrome among apprentice construction workers. American Journal of Industrial Medicine *42*(2):107-116.

Schoenmarklin RW, Marras WS, Leurgans SE [1994]. Industrial wrist motions and risk of cumulative trauma disorders in industry. Ergonomics *37*(9):1449-1459.

Welch LS, Hunting KL, Kellogg J [1995]. Work-related musculoskeletal symptoms among sheet metal workers. American Journal of Industrial Medicine *27*(6):783-791.

Tip Sheet #14. Ergonomic Hand Tools

Adapted from the booklet *Easy Ergonomics: A Guide to Selecting Non-Powered Hand Tools* (2004), a joint publication of the California Dept. of Occupational Safety and Health (Cal/OSHA) and NIOSH. Cincinnati, OH: U.S. Department of Health and Human Services, Centers for Disease Control and Prevention, National Institute for Occupational Safety and Health, DHHS (NIOSH) Publication No.2004-164. Other sources include:

Anton D, Cook TM, Rosecrance JC, Merlino LA [2003]. Method for quantitatively assessing physical risk factors during variable noncyclic work. Scandinavian Journal of Work Environment and Health *29*(5):354-362.

Dababneh A, Waters T [1999] The ergonomic use of hand tools: guidelines for the practitioner. Applied Occupational and Environmental Hygiene *14*:208-215.

Dababneh A, Lowe B, Krieg E, Kong YK, Waters T [2004]. A checklist for the ergonomic evaluation of nonpowered hand tools. Journal of Occupational and Environmental Hygiene *1*(12):D135-D145.

Merlino LA, Rosecrance JC, Anton D, Cook TM [2003]. Symptoms of musculoskeletal disorders among apprentice construction workers. Applied Occupational and Environmental Hygiene *18*(1):57-64.

Oregon Department of Consumer and Business Services, Workers' Compensation Division [2001]. Worksite modification digest. [wcd.oregon.gov//communications/publications/2184.pdf]. Date accessed: September 2005.

Radwin RG [2003]. Ergonomically-designed hand tools. Presentation at the American Industrial Hygiene Conference and Exposition. [homepages.cae.wisc.edu/~radwin/presentations.htm]. Date accessed: September 2005.

Tichauer ER, Gage H [1977]. Ergonomic principles basic to hand tool design. American Industrial Hygiene Association Journal *38*(11):622-634.

Tip Sheet #15. Easy-Hold Glove for Mud Pans

Moore JS [1997]. De Quervain's tenosynovitis: Stenosing tenosynovitis of the first dorsal compartment. Journal of Occupational and Environmental Medicine *39*(10):990-1002.

Rempel D, Keir PJ, Smutz WP, Hargens A [1997]. Effects of static fingertip loading on carpal tunnel pressure. Journal of Orthopaedic Research *15*(3):422-426.

Shaw G, Joyce T [2002]. Ergonomics of drywall finishing - How finishing tools and techniques affect repetitive strain injuries in the finishing trades. Conference Proceedings, 12th Annual Construction Safety and Health Conference, Rosemont, Illinois, May 21-23, 2002. [www.apla-tech.com/pdf/ergo.pdf]. Date accessed: September 2006.

Tip Sheet #16. Power Caulking Guns

Dababneh A, Lowe B, Krieg E, Kong YK, Waters T [2004]. A checklist for the ergonomic evaluation of nonpowered hand tools. Journal of Occupational and Environmental Hygiene *1*(12):D135-D145.

Methner MM [2000]. Identification of potential hazards associated with new residential construction. Applied Occupational and Environmental Hygiene *15*(2):189-192.

Tichauer ER, Gage H [1977]. Ergonomic principles basic to hand tool design. American Industrial Hygiene Association Journal *38*(11):622-634.

Tip Sheet #17. Reduced Vibration Power Tools

Griffin MJ, Howarth HVC, Pitts PM, Fischer S, Kaulbars U, Donati PM, Bereton PF [2005]. Guide to good practices on hand-arm vibration (V7.7). [www.humanvibration.com/EU/VIBGUIDE/HAV_Good_practice_Guide_V7.7_English_260506. pdf]. Date accessed: October 2006.

Naval Safety Center [2006]. Acquisition safety vibration. [www.safetycenter.navy.mil/acquisition/vibration/default.htm]. Date accessed: November 2006.

Tip Sheet #18. Power Cleaning and Reaming with a Brush

None.

Tip Sheet #19. Snips for Cutting Sheet Metal

Anton D, Rosecrance J, Gerr F, Reynolds J, Meyers A, Cook T [2007]. Effect of aviation snip design and task height on upper extremity muscular activity and wrist posture. Journal of Occupational & Environmental Hygiene *4*:99-113.

Merlino LA, Rosecrance JC, Anton D, Cook TM [2003]. Symptoms of musculoskeletal disorders among apprentice construction workers. Applied Occupational and Environmental Hygiene *18*(1):57-64.

Welch LS, Hunting KL, Kellogg J [1995]. Work-related musculoskeletal symptoms among sheet metal workers. American Journal of Industrial Medicine *27*(6):783-791.

Tip Sheet #20. Quick-Threading Lock Nuts

Pope DP, Silman AJ, Cherry NM, Pritchard C, Macfarlane GJ [2001]. Association of occupational physical demands and psychosocial working environment with disabling shoulder pain. Annals of the Rheumatic Diseases *60*:852-858

Sommerich CM, McGlothlin JD, Marras WS [1993]. Occupational risk factors associated with soft tissue disorders of the shoulder: a review of recent investigations in the literature. Ergonomics *36*(6):697-717.